M Harbison

Elements of Zoology

For schools and science classes

M Harbison

Elements of Zoology
For schools and science classes

ISBN/EAN: 9783337034740

Printed in Europe, USA, Canada, Australia, Japan

Cover: Foto ©Paul-Georg Meister /pixelio.de

More available books at **www.hansebooks.com**

Putnam's Elementary Science Series.

ELEMENTS OF ZOOLOGY:

FOR

SCHOOLS AND SCIENCE CLASSES.

BY

M. HARBISON,

HEAD MASTER, MODEL SCHOOL, NEWTOWNARDS.

NEW YORK:
G. P. PUTNAM'S SONS,
FOURTH AVENUE AND TWENTY-THIRD STREET.

PREFACE.

THE present Work is confined to a description of the Sub-kingdoms and Classes of Animals. As it is desirable that students should have some knowledge of Animal Physiology before entering on the study of Zoology, less space is devoted to the *Vertebrata* than is usual in Zoological works. The higher animals are described first, as it is believed that this arrangement is more suitable for beginners.

The Author has tried to adapt his language to the capacity of the students who attend classes in connection with the Department of Science and Art, and the more advanced pupils in schools. Technical terms have been freely used, but their derivation is always given, and full explanations will be found in the Glossary. Questions for Examination are added to each Chapter, which will, it is hoped, be found useful to both teachers and pupils.

Materials have been drawn from a great variety of sources, but the Author especially acknowledges his obligations to the works of PROFESSOR HUXLEY, DR. NICHOLSON, DR. HAUGHTON, and MR. GOSSE. The Illustrations, of which there are sixty-two, have been mainly derived from GERVAIS' *Élements de Zoologie*.

MODEL SCHOOL,
NEWTOWNARDS, *November, 1873.*

CONTENTS.

INTRODUCTION.

Definition—Differences between Organic and Inorganic Bodies—Classification, ... 9

CHAPTER I.
GENERAL CHARACTERS OF THE SUB-KINGDOMS.

Vertebrata — Annulosa — Annuloida — Mollusca — Molluscoida—Cœlenterata—Protozoa—Questions, . . . 14

CHAPTER II.
VERTEBRATA.

Mammalia — Aves — Reptilia—Amphibia—Pisces—Questions, ... 26

CHAPTER III.
ANNULOSA AND ANNULOIDA.

Arthropoda—Insecta—Myriapoda—Arachnida—Crustacea — Anarthropoda — Annelida — Chœtognatha — Annuloida—Scolecida—Rotifera—Turbellaria—Trematoda—Tœniada—Nematoidea—Acanthocephala—Gordiacea—Echinodermata—Questions, ... 58

CHAPTER IV.
MOLLUSCA AND MOLLUSCOIDA.

Cephalopoda — Pteropoda — Gasteropoda — Lamellibranchiata—Brachiopoda—Tunicata or Ascidioida—Polyzoa—Questions, ... 95

CHAPTER V.

Cœlenterata and Protozoa.

Actinozoa—Sea—Anemone—Coral—Coral Reefs—Ctenophora—Hydrozoa—Hydra—Corynidæ—Tubularia—Sertularidæ—Lucernaridæ—Medusæ—Infusoria—Rhizopoda—Amœba—Foraminifera—Radiolaria—Polycystinæ—Spongida—Gregarinida—Questions, . . 124

Glossary, ... 159

ZOOLOGY.

INTRODUCTION.

Natural History, strictly speaking, includes the study of all natural objects, whether animate or inanimate. It consists of two great divisions :—first, that which treats of inorganic or dead bodies ; second, that which deals with organic or living substances. The sciences which treat of inorganic bodies are termed Mineralogy and Geology, while that which deals with the properties common to organic beings (plants and animals) is termed *Biology* (Gr. *bios*, life ; *logos*, a discourse). This science comprises **Botany** (Gr. *botane*, a plant), which treats of the properties of plants, and **Zoology** (Gr. *zoon*, an animal; *logos*, a discourse), which investigates the nature and habits of animals. In the popular use of the term, Natural History is often synonymous with Zoology.

The most important differences between **organic** and **inorganic** bodies are :—

1. **Inorganic** bodies are either indefinite in shape *(amorphous)*, or assume regular forms called "crystals," which are bounded by plane surfaces and straight lines. **Organic** bodies are generally definite in shape, and are bounded by curved lines and rounded surfaces.

Inorganic bodies increase in size by the addition of similar particles to the outside. Organic bodies *grow* by the receiving and assimilation of matter into the interior.

2. Characteristic Differences between Plants and Animals.—Both plants and animals are organic bodies. They both possess life, and are able to propagate their kind. Although it is generally easy to distinguish between plants and animals, yet there is a border land in which it is difficult to determine where the one kingdom ends and the other begins. The following are the most characteristic differences of the two kingdoms; but, as will be seen, the distinctions can scarcely be looked upon as universal.

3. Chemical Composition.—Most of the characteristic substances found in plants, such as starch, sugar, &c., are composed of *three* elements—carbon, oxygen, and hydrogen. On the other hand, albumen and fibrine, so abundantly met with in animal tissues, are composed of *four* elements—carbon, oxygen, hydrogen, and nitrogen. This distinction, however, is not universal, as compounds of the three elements are met with in animals, while nitrogen is present in several vegetable products.

4. Feeling or Sensibility was selected by Linnæus as the leading characteristic of the animal kingdom. In such animals as possess a nervous system, the distinction is worthy of the position assigned to it by the great naturalist. But when it is remembered that two primary groups of animals, the Protozoa and the Cœlenterata, are utterly devoid of a nervous system, the characteristic fails to be universally applicable.

5. Locomotion.—This characteristic, which is so marked in the higher groups, also fails when we reach the more lowly organisms. Many of these, such as the sponges, corals, and corallines, are as much confined to a particular spot as plants are. On the other hand, many minute aquatic plants, such as the Diatomaceæ, are able to move about with great rapidity.

6. Organs of Digestion.—The possession of a receptacle, in which to deposit and digest their food, is a feature broadly distinctive of the animal kingdom. Many of the Protozoa, however, are destitute of any permanent alimentary apparatus, and require, when they take in nutriment, to extemporize a cavity for its reception.

7. **The Nature of the Food** is the most widely applicable of the distinctive features of the two kingdoms. This distinction may be briefly stated thus:—*Plants are able to convert inorganic matter into organic; but no animal, so far as our present knowledge extends, is able to subsist upon inorganic substances.* The food of plants consists principally of carbonic acid, water, and ammonia, together with small quantities of various salts. These substances are derived partly from the air, and partly from the earth. The carbon of the carbonic acid is retained by them, and the oxygen set free. Animals, on the other hand, deprive the air of its oxygen, and give off carbonic acid. Animals are able to subsist only on the complex organisms prepared for them by plants, the food of these being of a more simple character. "Plants are the great manufacturers in nature, animals the great consumers." (*Dr. Nicholson.*)

These distinctions, however, are not universally true. On the one hand, several of the *fungi* are as dependent on organic food as animals are; while, it is probable, that some microscopic animals may be able to subsist on inorganic compounds.

8. **Classification.**—Various systems of classification have been adopted by different naturalists. Lamarck divided the animal kingdom into two great divisions, the **Vertebrata** and **Invertebrata.** The first of these is still retained as a primary division, but the second contained such a heterogeneous assemblage that further division became necessary. Accordingly, Cuvier, another French naturalist, subdivided the Invertebrata into **Mollusca, Articulata,** and **Radiata.** As the new world of animal life revealed by the microscope could not be assigned to any of these groups, it became necessary to form a separate division, to which the name of **Protozoa** has been given.

Careful examination showed that Cuvier's Radiata

included animals with and without a nervous system, and differing widely in other important structures. This group has, therefore, been broken up, and its members assigned to various sub-kingdoms.

Following closely the arrangement proposed by Professor Huxley, we venture to divide the Animal Kingdom into seven primary groups or sub-kingdoms.

The following table gives the great divisions of the Animal Kingdom, together with the classes into which they have been subdivided:—

Sub-Kingdom I.—Vertebrata.

Class 1. **Mammalia.** Ex., Man, Horse, Deer.
" 2. **Aves** (Birds). Ex., Hawk, Crow, Parrot, Goose.
" 3. **Reptilia** (Reptiles). Ex., Crocodile, Turtle, Lizard, Snake.
" 4. **Amphibia.** Ex., Frog, Toad, Newt, Salamander.
, 5. **Pisces** (Fishes). Ex., Herring, Salmon, Shark, Ray.

Sub-Kingdom II.—Annulosa.

Division I.—Arthropoda.

Class 1. **Insecta.** Ex., House-Fly, Beetle, Bee, Butterfly.
" 2. **Myriapoda.** Ex., Centipede, Millipede.
" 3. **Arachnida.** Ex., Scorpion, Spider, Mite.
" 4. **Crustacea.** Ex., Lobster, Crab, Shrimp, Barnacle.

Division II.—Anarthropoda.

Class 1. **Chætognatha.** Ex., Sagitta.
" 2. **Annelida.** Ex., Earth-worm, Leech, Lobworm, Serpulæ.

Sub-Kingdom III.—Annuloida.

Class 1. **Scolecida.** Ex., Tape-worm, Fluke, Wheel-Animalcule.
  „   2. **Echinodermata.** Ex., Star-fish, Sea-Urchin, Encrinite.

Sub-Kingdom IV.—Mollusca.

Class 1. **Cephalopoda.** Ex., Nautilus, Cuttle-fish, Squid.
  „   2. **Pteropoda.** Ex., Clio Borealis, Dentalium.
  „   3. **Gasteropoda.** Ex., Snail, Limpet, Whelk.
  „   4. **Lamellibranchiata.** Ex., Mussel, Cockle, Scallop, Oyster.

Sub-Kingdom V.—Molluscoida.

Class 1. **Brachiopoda.** Ex., Terebratula, Lingula.
  „   2. **Tunicata** or **Ascidioida.** Ex., Sea-Squirt.
  „   3. **Polyzoa.** Ex., Flustra or Sea-Mat.

Sub-Kingdom VI.—Cœlenterata.

Class 1. **Actinozoa.** Ex., Coral, Sea-Anemone, Sea-Pen.
  „   2. **Hydrozoa.** Ex., Hydra, Sea-Fir, Sea-Jelly.

Sub-Kingdom VII.—Protozoa.

Class 1. **Infusoria.** Ex., Paramœcium, Vorticella.
  „   2. **Rhizopoda.** Ex., Sponge, Amœba, Foraminifera.
  „   3. **Gregarinida.** Ex., Gregarina.

CHAPTER I.

Characters of the Sub-Kingdoms.

Sub-Kingdom I.—Vertebrata.

9. The first primary division of animals is termed **Vertebrata**. It is so called on account of the almost universal presence of a vertebral column or backbone. The skeleton or hard parts are internal; some groups, however, such as the Crocodiles and Tortoises, have also an external skeleton. This skeleton consists of the bony case, called the cranium or skull, and a series of pieces, which are united together, and form the axis, termed the vertebral column. Attached to the vertebræ there is generally a number of bony arches, termed ribs, which are united by cartilages to the sternum or breastbone. Then there are the bones that support the limbs, which are greatly modified in the various groups.

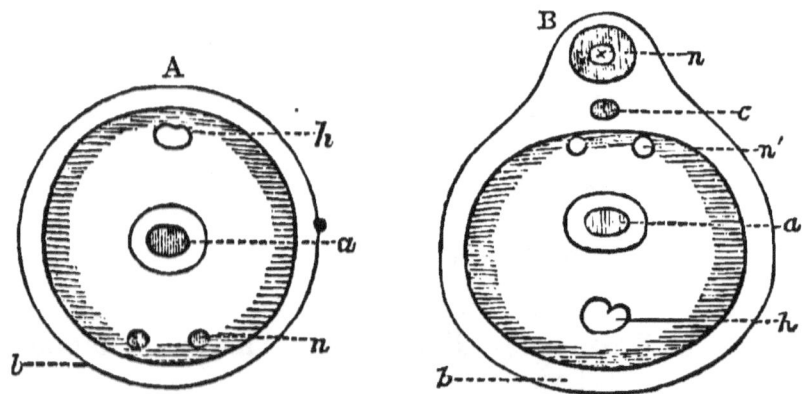

Fig. 1.

A, Transverse section of an invertebrate animal; B, Section of a vertebrate; *b*, wall of the body; *a*, alimentary canal; *h*, hæmal or blood system; *n*, principal nervous system; *n'*, sympathetic nervous system of vertebrata; *c*, notochord. It will be seen that, in the Vertebrata, the main mass of the nervous system is *dorsal* (Lat *dorsum*, the back), and *ventral* (Lat. *venter*, the belly) in the Invertebrata, while the blood system is ventral in the Vertebrata, and dorsal in the Invertebrata.

To the various parts of this skeleton the muscles are attached.

The nervous system of the Vertebrata is very characteristic. An invertebrate animal consists only of a single tube, which contains the nervous system as well as the organs of circulation and digestion. If a vertebrate animal, however, be cut in two, it will be seen to consist of two tubes, an upper and a lower. The upper tube is formed by two bony arches which proceed from the centre or body of the vertebra or joint of the backbone. These are termed the **neural arches** (Gr. *neuron*, a nerve), and form the canal in which the spinal cord is enclosed. This cord is connected with the brain, which is contained in the skull. These form the main mass of the nervous system.

The lower and much larger tube contains the alimentary canal and other digestive organs, the organs of circulation, and what is called the "sympathetic" nervous system which seems to correspond with the nervous system of the Invertebrata.

Some Vertebrata, as the serpents, &c., are destitute of limbs; and some, as the whales, possess only one pair. Generally, however, there are two pairs present, and no vertebrate animal has more than four limbs.

They have all a complete alimentary or intestinal canal provided with two openings, terminating respectively in a mouth and anus or vent. This canal is always completely shut off from the general cavity of the body. There is generally an œsophagus or gullet, and a stomach with large and small intestines.

The heart is formed of a variable number of cavities, called auricles and ventricles. Mammalia, or animals which nourish their young with a fluid called milk, and birds, have two auricles and two ventricles; reptiles and amphibia (frogs, toads, &c.), two auricles and one ventricle; fishes, one auricle and one ventricle.

The blood is red, with the exception of one species of fish. the lancelet, which has colourless blood. The colour is due to the globules called **corpuscles** (Lat. *corpus*, a body; *cle*, little), and not to the fluid in which they are suspended, this being colourless. It circulates to all parts of the body, in a system of close vessels, called veins and arteries. The arteries convey the blood from the heart, the veins carry it back.

There are always special respiratory or breathing organs for the purification of the blood. In mammals, birds, and reptiles, the oxygen required for this purpose is obtained directly from the atmosphere. Their breathing organs are called lungs. Fishes breathe by gills, and obtain the oxygen from the air dissolved in water. The amphibia breathe by gills when immature, and by lungs in the adult state. In all cases, the organs of respiration are connected with the commencement of the digestive tube.

In all the vertebrates the male and female organs are found in distinct individuals.

Divisions of the Vertebrata.

10. This sub-kingdom is divided into five classes:—
 1. **Mammalia** (warm-blooded quadrupeds which suckle their young).
 2. **Aves** (Birds).
 3. **Reptilia** (Reptiles).
 4. **Amphibia** (Frogs, Toads, &c.); and
 5. **Pisces** (Fishes).

11. The **Amphibia** and **Pisces** agree in possessing gills during either the whole or a part of life; the embryo is destitute of the membranous *sac*, termed the **amnion** (Gr. *amnos*, a lamb), and of the structure called the **allantois** (Gr. *allas*, a sausage); the red blood-corpuscles are oval in shape, and contain a nucleus. These two classes, according to Professor Huxley, form the division **Ichthyopsida**. (Gr. *ichthus*, a fish; *opsis*, appearance).

12. The **Reptilia** and **Aves** have also nucleated red blood-

corpuscles; the embryo is provided with *amnion* and *allantois;* gills are never present during any period of life; the skull is joined to the vertebral column by a single *condyle* (Gr. *condulos*, a knuckle) on the *occipital* bone; each half of the lower jaw consists of several pieces; the lower jaw is joined to the skull by a bone called the **quadrate** (Lat. *quadra*, a square) bone. These two classes form the division **Sauropsida** (Gr. *sauros*, a lizard).

13. The **Mammalia** form a distinct section. They agree with the **Sauropsida** in being destitute of gills, and in possessing an *amnion* and *allantois;* but differ from them in having circular red blood-corpuscles, without any nucleus, in possessing glands for the secretion of milk (**mammary glands**), and in being generally covered with hair. There are two *condyles* on the *occipital* bone, each half of the lower jaw consists of a single piece, and is united directly with the skull, and not to a quadrate bone.

Sub-Kingdom II.—Annulosa.

14. This sub-kingdom, which is also sometimes termed **Articulata,** comprises those animals that consist of a number of rings or segments, arranged behind one another in a straight line. There is no skeleton properly so called, but the hardness of the external covering of most of them compensates for the absence of the internal framework found in the **Vertebrata**. This hardness is sometimes due to lime, which, in crabs, lobsters, &c. (**Crustacea**) forms a hard external crust.

Sometimes the cuticle is hardened by a chemical substance termed **chitine** (Gr. *chiton*, a tunic), allied to horn, the skin, in such cases, somewhat resembling a tanned hide. This is the nature of the covering in the insects, centipedes, &c. To these hardened rings the muscles are attached.

There is always a nervous system which is situated below the alimentary canal. It generally consists of a pair of **ganglia** (Gr. *ganglion,* a knot) placed in each ring or segment, and connected by two nervous cords or **commissures** (Lat. *committo,* to join together), which run along the whole ventral surface of the body. The first or front pair of *ganglia,* is always situated *above* the alimentary

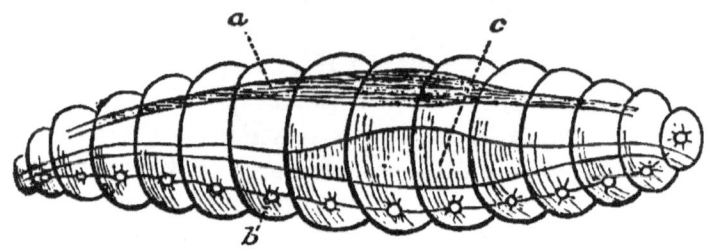

Fig. 2.—Diagram of an Annulose Animal.
a, hæmal or blood system; *b,* neural system; *c,* digestive system.

tube, and the nervous cords connecting this with the second pair pass round the gullet, and are in consequence termed the **œsophageal collar.** The nervous system is lodged in the same cavity as the digestive and circulatory organs, the body walls forming a single tube (see fig. 1).

The alimentary canal has always two openings, and is completely shut off from the cavity of the body.

Divisions of the Annulosa.

15. The sub-kingdom **Annulosa** comprises two great divisions—the **Arthropoda** (Gr. *arthros,* a joint; *podes,* feet), or animals with jointed limbs, and those in which these appendages are absent. The latter have been termed **Anarthropoda** (Gr. *an,* without).

The **Arthropoda** comprise :—
1. **Insecta** (Bees, Butterflies, &c.)
2. **Myriapoda** (Centipedes, &c.)
3. **Arachnida** (Spiders, &c.); and
4. **Crustacea** (Crabs, Lobsters, &c.)

The principal characteristic of this group is the possession of three or more pairs of jointed limbs composed of the chitinous substance described above. These limbs are sometimes confined to one region of the body, and sometimes distributed over all the segments of it.

The blood is a colourless fluid, containing corpuscles. The blood-vessels are situated *above* the alimentary canal.

Respiration is performed differently in the various classes.

The principal class in the division **Anarthropoda** is the **Annelida** or worms, which are destitute of jointed limbs.

Sub-Kingdom III.—Annuloida.

16. The members of this sub-kingdom were formerly included in the **Annulosa,** but have now been separated from it and erected into an independent group. The reasons assigned by Professor Huxley for this separation are :—Their bodies are not marked by definite rings or segments, there is no longitudinal chain of ganglia, and none of them possess limbs arranged in pairs.

Two classes have been assigned to this primary division,
1. **Echinodermata** (Gr. *echinos*, a hedgehog ; *derma*, skin), including sea-urchins, star-fishes, &c. ; and,
2. **Scolecida** (Gr. *scolex*, a worm), consisting principally of internal parasites.

At first sight there seems to be very little resemblance between these two classes,—a sea-urchin and a tape-worm being as unlike one another as possible. They both, however, possess what has been called the "water-vascular" system, that is, " a series of canals communicating with the exterior by means of apertures situated upon the surface of the body, and branching out more or less extensively into its substance."

Sub-Kingdom IV.—Mollusca.

17. The **Mollusca** (Lat. *mollis*, soft,) are invertebrated animals whose bodies are covered by a soft skin, which envelops them like a loose sac. This covering is called the *mantle*. It secretes the shell which most of them possess, and to it the muscles are attached. This shell is

composed of carbonate of lime, and sometimes consists of a single piece or "valve," sometimes of two valves united together by a hinge.

The digestive tube is always completely separated from the general cavity of the body, and is provided with an anal opening.

The higher groups possess an organ situated at the entrance to the alimentary canal, which is commonly called the "tongue;" but which, on account of the minute flinty teeth with which it is furnished, Professor Huxley has termed an *odontophore* (Gr. *odous*, a tooth; *phero*, I carry), or tooth-bearer. It works backwards and forwards on a cartilaginous cushion like a chain saw. It is by means of this *odontophore* that the slugs and snails make such havoc in our gardens.

The blood in this group is colourless. There is a distinct heart, with two cavities, which occupies the dorsal position. The blood passes from the gills, or other breathing organ, to the *auricle*, thence to the *ventricle*, which propels it through the body.

The nervous system is very distinctive. It consists of three pairs of ganglia united by nervous cords. One of these pairs, situated in front above the œsophagus, is termed the *cerebral ganglia;* another, which sends branches to the foot, is termed the *pedal ganglia;* and the third,

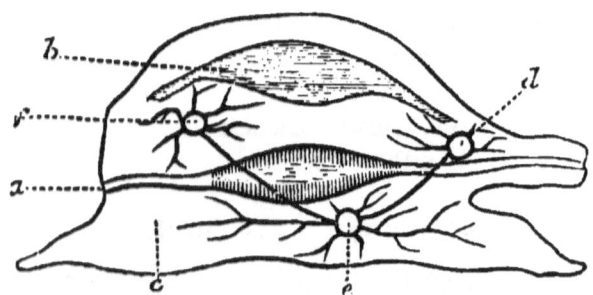

FIG. 8.—DIAGRAM OF A MOLLUSC.

a, digestive tube; *b*, heart; *c*, foot; *d*, cerebral ganglia; *e*, pedal ganglia; *f*, parieto-splanchnic ganglia.

which communicates with the walls of the body, as well as with the digestive and reproductive organs, is termed the *parieto-splanchnic ganglia* (Gr. *parietes*, walls; *splanchna*, the internal organs).

Respiration is generally effected by gills or branchiæ. In the land snails, and in some fresh water molluscs, the breathing organ is termed a *pulmonary sac*. It consists of an opening hollowed out on the right side of the body, near the neck, and supplied with blood-vessels. The blood in these animals is purified by coming directly into contact with the oxygen of the atmosphere, and not through the medium of water as in those that breathe by gills.

18. The sub-kingdom **Mollusca** is divided into the following classes, viz.:—

1. **Cephalopoda** (Gr. *cephale*, the head; *poda*, feet).
2. **Pteropoda** (Gr. *pteron*, a wing; *poda*, feet).
3. **Gasteropoda** (Gr. *gaster*, the stomach).
4. **Lamellibranchiata** (Lat. *lamella*, a plate; Gr. *branchia*, a gill).

Sub-Kingdom V.—**Molluscoida.**

19. This division is allied to the **Mollusca**, but its characters are sufficiently distinct to entitle it to rank as an independent group.

The heart, when present, is in the form of a simple tube or sac, and is never separated into auricle and ventricle.

The nervous system generally consists of a single ganglion.

The mouth is either surrounded with ciliated tentacles, or with long arms, which have a ciliated fringe.

The alimentary tube has two openings, and is completely shut off from the cavity of the body.

There are three classes in this sub-kingdom :—

1. **Brachiopoda** (Gr. *brachia*, arms; *poda*, feet).
2. **Tunicata** or **Ascidioida** (Lat. *tunica*, a cloak; Gr. *askos*, a bottle; *eidos*, form).
3. **Polyzoa** (Gr. *polus*, many; *zoon*, an animal).

Sub-Kingdom VI.—Cœlenterata.

20. The **Cœlenterata** (Gr. *koilos*, hollow; *enteron*, an intestine) form a very distinct division of the animal kingdom. The red and white corals, many of the corallines, sea jellies, &c., belong to it. These were placed by Cuvier, along with some other groups, in the division named by him **Radiata**. From their resemblance to plants many of them have also been called **Zoophytes**. (Gr. *zoon*, an animal; *phuton*, a plant).

The body wall always consists of two layers, an outer and an inner. The external covering is called the **ectoderm** (Gr. *ectos*, outside; *derma*, skin); the inner lining, the **endoderm** (Gr. *endos*, within). These layers have a cellular structure, and are generally furnished with **cilia** (Lat. *cilium*, an eyelash), which constantly vibrate.

There is a circle of tubular arms round the mouth. These are called **tentacles** (Lat. *tento*, I touch), and are used by the animals in seizing their prey.

The most distinctive feature in this group is the digestive apparatus. In all the divisions hitherto described it was seen that the digestive canal is completely shut off from the body cavity, and communicates with the exterior by an **anus** or vent. In the **cœlenterata**, however, the alimentary tube opens at its inner end into the general cavity of the body, and communicates freely with it. As there is no anal opening, the indigestible portions of the food are expelled from the mouth.

Another peculiar feature of the group is the possession of stinging organs in the form of "thread cells." These may be examined by the microscope in the common freshwater *hydra*. "These cells are oval elastic sacs, containing a long coiled filament, barbed at its base, and serrated along the edges. When fully developed the sacs are filled with fluid, and the slightest touch is sufficient to cause the retroversion of the filament, which then projects beyond the sac for a distance which is not uncommonly equal to many times the length of the latter." (*Huxley.*) These organs are doubtless used by the animals in killing their prey, as well as in defence. Some of the "jelly fishes" have filaments large enough to penetrate the human skin.

No organs of circulation or respiration are known to exist. Only in one group is there any trace of a nervous system. All of them have distinct reproductive organs.

This sub-kingdom is divided into two classes:—

1. **Actinozoa** (Gr. *aktis*, a ray; *zoon*, an animal).
2. **Hydrozoa** (Gr. *hudra*, a water serpent; *zoon*, an animal).

SUB-KINGDOM VII.—**Protozoa.**

21. The **Protozoa** (Gr. *protos*, first; *zoon*, an animal), as the name signifies, is the lowest primary division of the animal kingdom. Most of the animals in this division are so small that they can only be seen by a powerful microscope. Some, however, as the sponges, are of large size. On account of their minute size they are popularly called **animalcules**. Until recently, they were comparatively unknown even to the scientific world. Most people are still unaware of their existence. They are essentially aquatic animals, and may be found abundantly in any stagnant pond, or in water containing decayed animal or vegetable matter.

The bodies of the **Protozoa** are composed of a substance termed **sarcode** (Gr. *sarx*, flesh; *eidos*, form). It is also sometimes called **protoplasm** (Gr. *protos*, first; *plasso*, I mould.) This substance is jelly-like, and has been described by some observers as resembling the white of an egg; by others it is said to be like a particle of "thin glue." It is mainly composed of albumen, and is without any definite structure.

Except in one group, the Infusoria, there is no permanent mouth nor any digestive organ whatever. No trace of a nervous system has been observed.

In the **Infusoria** locomotion is effected by cilia, which are kept perpetually in motion. The lower forms have the power of extemporizing processes termed **pseudopodia** (Gr. *pseudos,* false; *poda,* feet). These are sometimes finger-like, and sometimes fine filaments resembling minute threads.

This sub-kingdom has been divided into three classes:—

1. **Infusoria** (so called because they are met with in *infusions* of vegetable matter).
2. **Rhizopoda** (Gr. *rhiza,* root; *poda,* feet).
3. **Gregarinida** (Lat. *gregarius,* occurring in flocks).

QUESTIONS.—I.

1. Strictly speaking, what sciences are included in "Natural History"?
2. What is the popular use of this term?
3. What are the two great divisions of natural objects?
4. What is Biology, and what sciences does it include?
5. Define Zoology.
6. What are the leading distinctions between organic and inorganic bodies?
7. What are the principal differences between plants and animals?
8. What substances do living animals and plants respectively contribute to the atmosphere?
9. Name the primary groups of animals, and the classes into which they are subdivided.
10. What are the principal characters of the sub-kingdom *Vertebrata?*
11. What are the principal parts of the vertebrate skeleton?
12. Describe the nervous system.
13. How are the brain and spinal cord protected?
14. How does a transverse section of a vertebrate animal differ from a similar section of an invertebrate?
15. How many limbs have the *Vertebrata,* and what are the exceptions to this rule?
16. How does the heart differ in the various groups?
17. What is the nature of the blood?
18. What are "corpuscles"?
19. What vertebrate animal has colourless blood?

CHARACTERS OF THE SUB-KINGDOMS.

20. How do the breathing organs differ in the various classes?
21. What are the primary divisions of the *Vertebrata?*
22. Name the classes or subdivisions.
23. What are the points of agreement between *Amphibia* and Fishes?
24. How do *Reptilia* and Birds resemble one another?
25. What are the leading distinctions of the *Mammalia?*
26. Define the sub-kingdom *Annulosa.*
27. How does the skeleton differ from that of the vertebrates?
28. What is "chitine"?
29. Describe the nervous system.
30. What are "ganglia"?
31. What are "commissures"?
32. Name the primary divisions of the *Annulosa.*
33. Name the sub-divisions, and give examples.
34. What is the nature of the blood in the *Arthropoda?*
35. Why have the *Annuloida* been separated from the *Annulosa?*
36. What are the principal characteristics of this sub-kingdom?
37. What is the "water vascular system"?
38. What classes are included in the *Annuloida?*
39. State why these classes have been placed in the same sub-kingdom.
40. Why are the *Mollusca* so called?
41. What classes are included in this sub-kingdom?
42. Describe the digestive apparatus in the *Mollusca.*
43. Describe the circulatory organs.
44. Describe the nervous system.
45. How is respiration effected in this group?
46. What is an "odontophore"?
47. Define the *Molluscoida.*
48. What classes are contained in this sub-kingdom?
49. Why have they been separated from the *Mollusca?*
50. Define the *Cœlenterata,* and give the derivation of the word.
51. What is the peculiarity of the digestive apparatus in this sub-kingdom?
52. What classes does it embrace?
53. What is the derivation of *Protozoa?*
54. Name the classes included in this sub-kingdom.
55. What is the general character of these groups of animals?
56. What is "sarcode"?
57. What are "cilia"?
58. What are "pseudopodia"?

CHAPTER II.

Vertebrata.

Class I. Mammalia.

22. The highest class of vertebrate animals is termed *Mammalia* (Lat. *mamma*, the breast). It is so called because the young, being brought forth in a state more or less helpless, are, for a time, nourished by a fluid called milk, secreted by a set of glands termed "mammary glands."

The following are the most important characteristics of the Mammalia:—

They are covered with hair; the skull is united to the vertebral column by two occipital condyles; each half of the lower jaw consists of a single piece; the lower jaw is joined directly with the skull, and not to the quadrate bone; the heart has four cavities—two auricles and two ventricles; the red blood-corpuscles are circular, and non-nucleated; the blood is hot; respiration is always effected by lungs; the diaphragm completely separates the cavities of the thorax and abdomen; the hemispheres of the brain are united by a *corpus callosum;* the embryo is provided with an *amnion* and an *allantois;* the young are nourished by milk, secreted by mammary glands.

23. **Skeleton.**—The **skull** in the **Mammalia** is united to the vertebral column by two condyles on the occipital bone. There are two corresponding "facets" on the **atlas**, or upper joint of the vertebral column. Each half of the lower jaw consists of only a single piece, and is articulated directly with the skull, and not to the **quadrate** bone, as in birds and reptiles.

24. The divisions of the **vertebral column** usually recognized are:—

1. The **cervical** (Lat. *cervix*, the neck) region or neck.

VERTEBRATA—MAMMALIA.

The number of vertebræ in this division is generally *seven*, whether the neck be long, as in the giraffe, or short, as in the whales. The only exceptions to this rule are, the three-toed sloth, which has *nine* neck bones; and the manatee, which has *six*.

2. The **dorsal** (Lat. *dorsum*, the back) region. The number of vertebræ in this region varies from *ten* to *fifteen* or more. The usual number is twelve or thirteen. To the dorsal vertebræ the *true ribs* are attached. These enclose the cavity of the chest, and are united in front by cartilages to the **sternum** or breast bone.

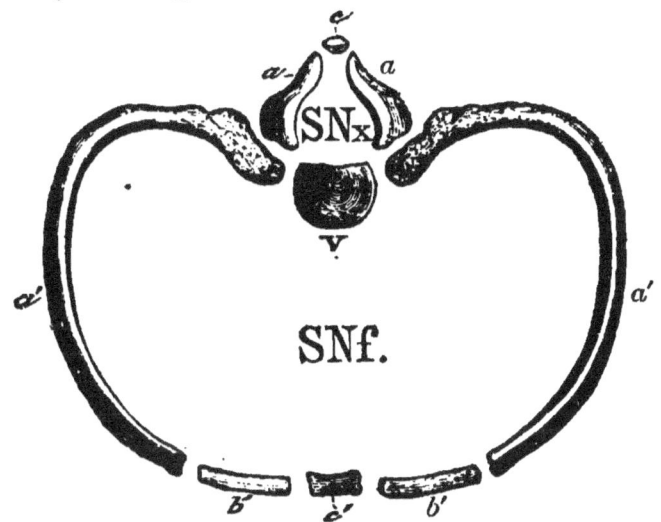

Fig. 4.—Dorsal Vertebra with Attached Ribs.

aa, neural arches; *snx*, neural canal; *v*, vertebral centre; *a'a'*, ribs enclosing the visceral cavity, *snf*; *b'b'*, cartilages; *c'*, sternum.

3. The **lumbar** (Lat. *lumbus*, a loin) vertebræ occupy the region of the loins. They vary in number from *four* to *seven*. There are four in man.

4. The **sacral** bones which follow vary from *one* to *nine* in number. They are *anchylosed*, or immovably joined together, so as to form a single bone.

5. The **caudal** (Lat. *cauda*, the tail) or tail vertebræ range from *four* to *forty-six*. They move more freely upon one another than the other vertebræ.

25. There are usually *four limbs* present, but the whale tribe have only the front pair. On this account the mammalia are often called **quadrupeds.** This term, however, would also include the *amphibia* and most reptiles.

The *fore limb* is joined to the trunk by the **scapula** or shoulder blade, and the **clavicle** or collar bone. The upper part of the limb consists of a single bone, called the **humerus**; the lower part of two bones, the **radius** and **ulna**; then follow the **carpal** or wrist bones, the **metacarpal** bones, which form the upper part of the hand, and the **phalanges**, or bones of the fingers or toes.

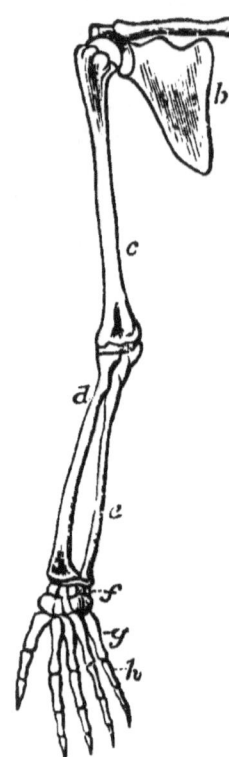

FIG 5.—FORE LIMB OF A MAN.

a, clavicle or collar-bone; *b*, scapula or shoulder-blade; *c*, humerus; *d*, radius; *e*, ulna; *f*, carpal or wrist-bones; *g*, metacarpus or bones of the hand; *h*, phalanges or finger-bones.

The parts of the *hind limb* are similar to those just mentioned; but they are called by different names. The hind limb is joined to the trunk by a number of bones, which are generally anchylosed together, and termed the **innominate** bone. To this is joined the **femur** or thigh bone; then the leg bones, **tibia** and **fibula**; the **tarsus** or ankle bones; the **metatarsus**, or bones of the foot; and the *phalanges* of the toes. There are usually five digits, with three phalanges in each. The horse, however, has only one digit, corresponding to the middle finger. The ox has bones corresponding to two fingers; and there are other variations from the rule found in some of the carnivora.

26. Digestion.—Except the Baleen whale and the great ant eater, all mammals possess *teeth*, which are inserted in distinct sockets in the jaws. They have generally two sets of teeth; the first, termed the *milk* or *deciduous* teeth, being succeeded by the *permanent* teeth. Generally, there are four different sorts of teeth present—**incisors, canines, præmolars,** and **molars.** The exceptions to this rule are numerous, the number and varieties of teeth present differing much in the various "orders" into which the class Mammalia is divided.

The food during mastication is mixed with a substance called **saliva,** secreted by the salivary glands. These glands are absent in the whales. It then passes through a tube called the **œsophagus** or gullet into the stomach, where it is acted upon by a fluid termed the **gastric juice.** By means of this fluid it is converted into a substance called **chyme.** This chyme passes through the **small intestine,** where it is acted upon by the **bile,** secreted by the **liver;** and the **pancreatic juice,** produced by another gland called the **pancreas.**

"The special function of the salivary glands is to digest the *starch* compounds of the food; and the special function of the gastric juice is to digest the *nitrogenous* elements of the food; the remaining or *fatty* elements of food are, in like manner, digested, and prepared for assimilation by the pancreatic juice." (*Dr. Haughton.*)

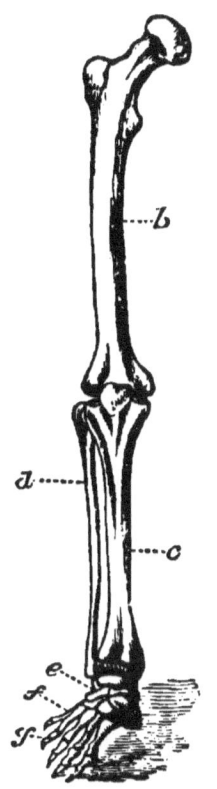

FIG. 6.—HIND LIMB OF A MAN.

b, femur or thigh bone; *c*, tibia; *d*, fibula; *e*, tarsus or ankle bones; *f*, metatarsus; *g*, phalanges or bones of the toes.

By these fluids the chyme is changed into a milky substance termed **chyle,** which is conveyed to the blood-vessels by a set of tubes called **lacteals.**

The small intestine is followed by a wider tube termed the **large intestine,** which is provided at the outer extremity by an opening called the **anus**. Through this tube the indigestible portions of the food are carried off as excrement.

27. Circulation.—The heart, which is surrounded by a membranous sac called the **pericardium**, contains four chambers. The right side consists of two cavities, the right auricle and the right ventricle; the left side has two similar cavities, the left auricle and the left ventricle. The blood passes from the auricles to the ventricles; but, the apertures being guarded by valves, it cannot pass from the ventricles to the auricles. There is no communication between the right and left sides of the heart.

Two great veins, the superior and inferior **venæ cavæ**, receive the impure or **venous** blood from the entire body, and pour it into the right auricle, from which it passes to the right ventricle. From this chamber it is driven to the lungs through the **pulmonary artery,** where it is changed by the oxygen of the air into **arterial** blood. The **pulmonary veins** convey it back to the left auricle, from which it passes to the left ventricle, and is by it propelled through the **aorta** and its numerous branches to the various parts of the body.

The greater number of the corpuscles are red. They are circular in form, except in the camels, and are destitute of any solid particle or nucleus. The blood is hot.

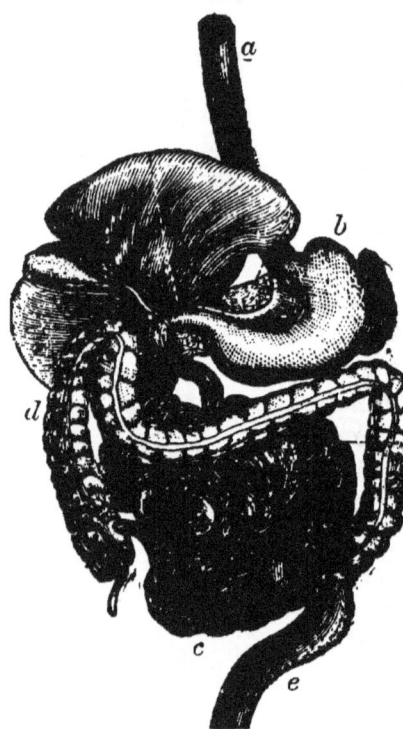

FIG. 7.--DIGESTIVE SYSTEM OF A MAMMAL.

a, œsophagus; *b*, stomach; *c*, small intestine; *d*, large intestine; *e*, rectum.

28. Respiration.—The **thorax** or chest is separated from the **abdomen** by a muscular partition called the **diaphragm**, which plays an important part in the work of respiration. In the abdomen are placed the greater part of the alimentary canal, the stomach, the liver, kidneys, &c. The heart and lungs are situated in the thorax. There are two lungs, which are spongy and elastic, and abound in air cells. They are freely suspended in the cavity of the thorax, and each of them is enclosed in a membranous sac called the **pleura** (Gr. *pleura*, the side). Their **bronchi**, or branches of the windpipe which traverse them, never communicate with air sacs in the body, as those of birds do.

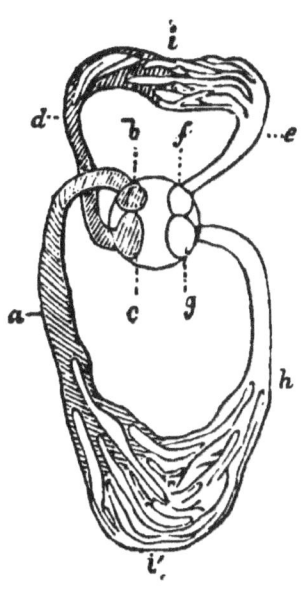

FIG. 8.—CIRCULATION OF A MAMMAL.

a, vena cava; *b*, right auricle; *c*, right ventricle; *d*, pulmonary artery; *e*, pulmonary veins; *f*, left auricle; *g*, left ventricle; *h*, aorta; *i*, capillaries of the lungs; *i'*, capillaries of the system. The venous system is shaded.

29. Nervous System.—The brain in the Mammalia is much larger than in the other vertebrates. The two hemispheres of the **cerebrum**, or principal mass of the brain, are united by a band of nerve fibres termed the **corpus callosum** (Lat. "the firm body").

30. Sight and Hearing.—All mammals possess eyes, but in some instances the sight is very imperfect. Except the whales, and some of the seals, all have an external ear.

31. Covering.—The covering consists of **hair**, which is only found in this class. Hair is of the same nature as feathers, but does not split up as feathers do. It is always developed by some part of the skin, except in the whales and the scaly ant-eater. In the hedgehog and porcupine the hairs take the form of spines. The armadilloes are partly covered with bony plates.

32. The members of this group are usually arranged in two* great divisions or sub-classes :—

1. **Placental mammals.**
2. **Non-placental mammals.**

The **placenta** (Lat. *placenta*, a cake) is a structure developed by the **allantois**, and composed of vascular tissue. The young of mammals, possessing this placenta, remain for a considerable period within the body of the mother, where they are nourished by arterial blood supplied through the minute vessels of this organ. In the non-placental mammals the young are brought forth much earlier; and, not having been nourished by a placenta, are born in a very imperfect condition. They are then placed by the mother within a "pouch" beneath the abdomen, where they become attached to the nipple; and, being unable to obtain the milk by suction, it is forced into their mouths by a special muscle connected with the mammary gland.

Class II.—Aves.

33. Aves (Lat. *avis*, a bird) or birds form the second class of vertebrate animals.

As already stated, birds have been placed along with reptiles in the section of the Vertebrata which has been termed **Sauropsida**. The points of agreement are :—At no period of life are gills present; the skull is joined to the vertebral column by a single condyle on the occipital bone; the lower jaw is composed of several pieces, and is united to the skull by a quadrate bone; the red blood-

---

* Professor Huxley, following De Blainville, divides mammals into three sub-classes :—

1. Ornithodelphia, which embraces the order *Monotremata*.
2. Didelphia, which corresponds to the *Marsupialia*.
3. Monodelphia, which includes all other mammals. (See Huxley's *Classification of Animals.*)

corpuscles are oval and nucleated; the embryo is provided with an *amnion* and an *allantois*.

34. Skeleton.—In order to adapt birds for flight, the skeleton is remarkable for its combined lightness and strength. There is a larger proportion of phosphate of lime present than in the bones of mammals; and in adult birds, many of the bones are filled with air instead of marrow.

In full grown birds, the various bones that form the skull, are amalgamated into a single piece. There is only a single condyle on the occipital bone, as in the reptiles. In young birds, each half of the lower jaw consists of six pieces; but these, in the mature animal, are so amalgamated, that the lines of junction cannot be distinguished. The lower jaw is joined to the quadrate bone, and not directly to the skull, as in the mammalia.

The neck is long and flexible, and consists of from nine to twenty-four vertebræ. As the beak is the only organ of prehension possessed by these animals, such a neck is required to enable them to procure their food. It is also long enough to reach a gland situated at the base of the tail, where the oil with which they trim their feathers is secreted.

There are from six to ten *dorsal* vertebræ, several of which are anchylosed together. By this arrangement, that part of the vertebral column is made sufficiently strong to bear the strain caused by the motion of the wings in flight. In birds, however, that do not fly, such as the ostrich tribe and the penguins, the dorsal vertebræ are movable upon each other.

The vertebræ between the dorsal and caudal regions of the spine are amalgamated into a single bone called the **sacrum**, which is anchylosed with the **innominate** bones to form the **pelvic arch.**

The innominate bones are always anchylosed together, but are never united below, except in the ostrich.

The **caudal** vertebræ, eight to ten in number, are more or less movable. The last joint, which consists of several vertebræ anchylosed together, is, from its shape, termed the "ploughshare bone," and is nearly perpendicular to the vertebral column. It supports the large quill feathers of the tail. In an extinct bird, the **archæopteryx** (Gr. *archaios*, ancient; *pteryx*, a wing), whose remains are found in the oolitic rocks, there was no ploughshare bone, and the tail consisted of twenty distinct vertebræ, each of which bore a pair of quill feathers.

The **sternum** is very large, and in flying birds is provided with a ridge or "keel." To the *sternum*, the muscles which move the wings are attached. The keel is absent in the **Cursores**, an order which includes the ostrich, emeu, cassowary, &c. The number of **ribs** varies from seven to eleven pairs.

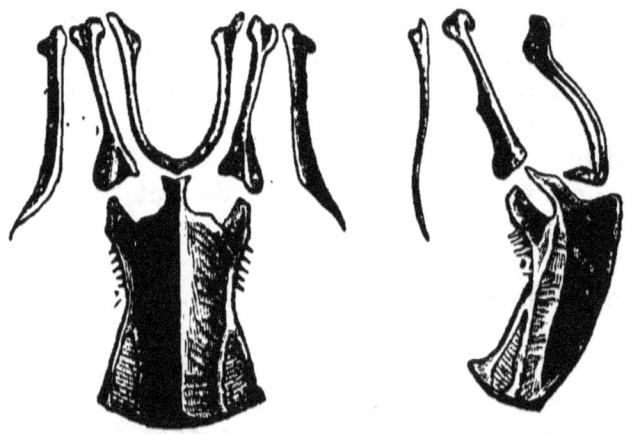

Fig. 9.—Sternum, Furculum, Scapulæ, and Coracoid Bones of a Sparrow.

The **scapula, clavicle,** and **coracoid** bone form what is called the "pectoral arch." The clavicles are united in front, forming a V-shaped bone, which is termed the **furculum** (Lat. *furca*, a fork) or "merrythought." The **coracoid** (Gr. *korax*, a crow) bone corresponds to the "coracoid process" in man, but in birds is a distinct bone. It is united at one end to the clavicle and scapula, at the other to the sternum.

Of the **fore-limb** or wing, the upper bone is the

humerus, which is united to the pectoral arch, and is followed by the **radius** and **ulna**. The radius is very slender. Then follow two small **carpal** bones, and two **metacarpal** bones, which are anchylosed at both extremities. There is a rudimentary thumb situated at the upper end of the metacarpal bones. This thumb consists of a single joint, and carries the "bastard wing." The metacarpal bones support two "fingers," which correspond to the index and middle fingers of the human hand. One of the fingers consists of a single phalanx or joint; the other has two or three.

FIG. 10.—WING OF A SPARROW.

With regard to the structure of the **hind limb**, there is a very short **femur** united to the "pelvic arch." This is followed by the **tibia**, to which a very thin **fibula** is anchylosed. The upper part of the **tarsus** is amalgamated with the tibia, and the lower portion with the metatarsus, forming the bone called the **tarso-metatarsus**. In the waders this bone is very long.

There are usually four **toes**, but some domestic birds possess a fifth toe. Three of the toes are commonly placed before, and one behind. One group, however, has two before and one behind. The swifts have the whole four directed forwards. In several of the swimmers, and

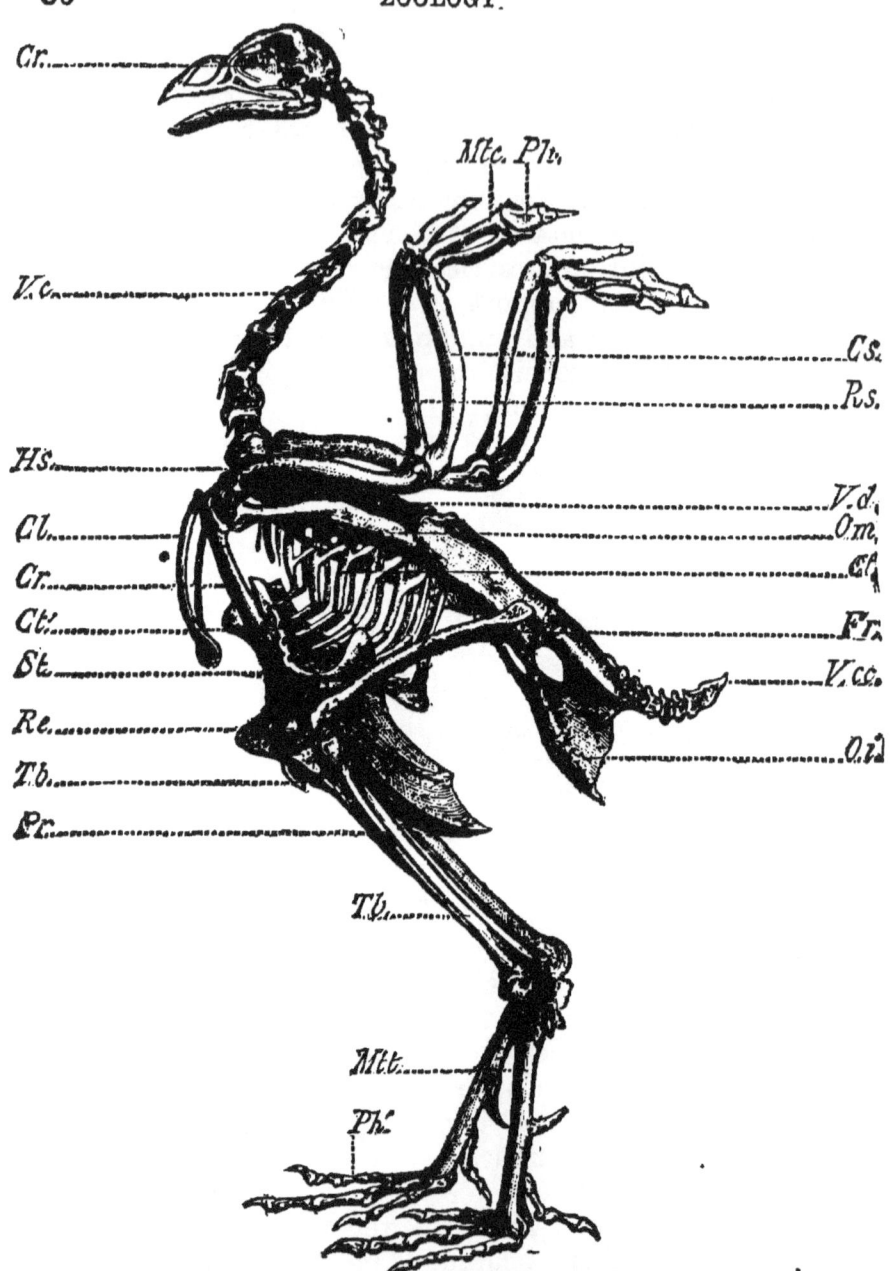

Fig. 11.—Skeleton of Cock.

c r, the cranium; *v c*, cervical vertebræ; *c l*, the furculum or merrythought; *c r*, coracoid bone; *s t*, sternum; *v d*, dorsal vertebræ; *o i*, innominate bone; *v c c*, caudal vertebræ; *h s*, humerus; *c s*, ulna; *r s*, radius; *m t c*, metacarpus; *p h*, phalanges; *f r*, femur; *t b*, tibia; *p r*, fibula; *m t t*, tarso metatarsus; *ph'*, phalanges of the toes.

in the emeu and cassowary, the hind toe is wanting or rudimentary. The African ostrich has only two toes. The hind toe, when present, has two, the innermost three, the next four, and the third five phalanges.

35. Digestion.—Birds are destitute of teeth. The **mandibles** or jaws are sheathed with a horny substance forming a beak. They are provided with sali**vary glands**, which are much smaller than in mammals. The food passes through a long gullet into a pouch called the **crop**, where it remains for some time. This crop, which may be considered an expansion of the gullet, is wanting in some birds. It is then transferred to the true digestive stomach, where it is mixed with the gastric juice. Having been imperfectly masticated, owing to the absence of teeth, the food, after leaving the true stomach, is conveyed to a muscular bag, called the **gizzard**, provided with a rough horny lining, which, assisted by gravel swallowed by the bird, reduces it to a pulpy mass. This kind of gizzard is found in grain-eating birds only. In birds of prey which live upon flesh its coats are thinner, and it is a much less

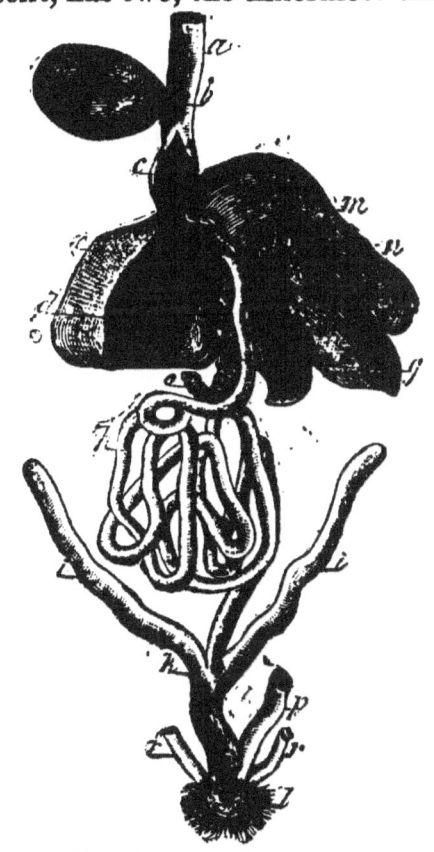

FIG. 12.—DIGESTIVE APPARATUS OF THE COMMON HEN.

a, the lower part of the œsophagus; *b*, crop; *c*, true digestive stomach; *d*, gizzard; *e*, muscular wall of gizzard; *g*, duodenum; *h*, small intestine; *ii*, intestinal *cæca*; *k*, commencement of large intestine; *l*, cloaca; *m*, liver; *o*, pancreas.

powerful organ. The food, after leaving the gizzard, passes into the small intestine, where it is acted upon by the bile and pancreatic juice. The large intestine, which is very short, passes into a bladder-like cavity called the **cloaca** (Lat. *cloaca*, a sink), which also receives the secretions of the kidneys and the genital organs. There is a similar arrangement in reptiles.

36. Circulation.—The mean temperature of the blood of birds is 110°, while that of mammals is about 100°. The higher temperature is partly due to the great activity of these animals, and partly to the "double respiration," by which the blood is aerated in the air sacs of the body, as well as in the lungs. The covering of feathers, which are bad conductors of heat, assists in keeping up the high temperature.

The red corpuscles of the blood are oval in shape, and contain a central solid particle or nucleus.

The structure of the heart and the nature of the circulation in birds and mammals are very much alike. The heart has four chambers, two auricles and two ventricles; and the right and left sides have no communication with each other. The venous blood is received from the body into the right auricle. It then passes into the right ventricle, which sends it to the lungs, through the pulmonary artery. Having been aerated in these organs, and changed into arterial blood, it is conveyed by the pulmonary veins to the left auricle, thence to the left ventricle, which drives it through the aorta and its branches to all parts of the body.

The valve which separates the right auricle from the right ventricle, instead of being "tricuspid" as in mammals, is of a triangular form. The action of the heart is more rapid than in mammals, and its walls and valves are relatively stronger.

37. Respiration.—Birds have two lungs of a bright red colour, and spongy structure. They are placed at the back of the chest, and are not suspended in a pleural membrane, as in mammals.

Fig. 13.—Respiratory Apparatus of the Hen.

b, trachea; *c*, bronchus; *d*, spongy part of lungs; *e*, air sac situated in the breast; *f*, air sacs in the region of the shoulders; *g* and *h*, large air sacs placed in the abdomen.

The cavities of the chest and abdomen are not separated by a diaphragm, this partition being only rudimentary.

The bronchial tubes communicate by a series of openings on the surface of the lungs with a number of air sacs situated partly in the chest, and partly in the abdomen. These sacs are connected with the interior of the bones, which also contain air cells. As has been already stated, the blood is aerated in these air-receptacles as well as in the lungs. By filling these sacs with air, the animal is enabled to reduce its specific gravity at will.

There are no air-cells in the bones of the penguin and some other aquatic fowls; and only a few of the bones in the ostrich tribe are "pneumatic."

38. Nervous System.—The brain is relatively smaller than in mammals, and is without convolutions. The two hemispheres of the **cerebrum** are not united by a *corpus callosum*.

39. Senses.—The *eyes* of birds are in front, of a conical shape, and possess a *third* eyelid. The sense of sight is always very acute. There is no external *ear*. The *nostrils* are placed on the sides of the upper mandible. They are sometimes surrounded by bristles, and are sometimes covered by a cartilaginous scale.

40. Covering.—Feathers form the peculiar covering of birds. Being non-conductors of heat, they assist in keeping up the high temperature of the body. A feather consists of several distinct parts. First, there is a horny tube, called the **quill**; then the **shaft**, which has a horny surface, with a groove on the under side; the interior being filled with a pith-like substance. The **webs** are situated at the sides of the shaft. The divisions of the web are called **barbs**, each of which is covered with a number of **barbules**. Generally, the barbs are kept together by the barbules, which are provided with hooks for this purpose. In the ostrich tribe, however, the barbs are disconnected.

There is an **oil gland** at the base of the tail, with the

secretions from which they anoint their feathers, and thus render them impervious to moisture.

41. Development.—Birds are **oviparous** (Lat. *ovum*, an egg; *pario*, I bring forth) animals. The ovary is situated on the left side. During the passage of the egg through the oviduct, it receives, first, the "white" or albuminous covering with which the embryo is nourished; and, afterwards, the calcareous shell, with which it is covered. From the oviduct it reaches the cloaca, and passes thence to the outer world. It is then subjected for a longer or shorter period to a process of **incubation**, by which the chick is fully developed. To enable it to break the shell, the bill of the young bird is provided with a protuberance of calcareous matter, which afterwards disappears.

Those birds which have been termed **Autophagi** (Gr. *autos*, self; *phago*, I eat) are able to run about and provide for themselves as soon as they leave the shell. The other section, **Heterophagi** (Gr. *heteros*, other; *phago*, I eat), come from the egg, blind and naked; and require to be protected from cold, and fed by the parent birds for a considerable time.

Class III.—Reptilia.

42. The **Reptilia** (Lat. *repto*, I creep) form the third class of vertebrate animals, and the second of the section, **Sauropsida.** This class includes crocodiles, lizards, serpents, turtles, and tortoises.

They possess the following characters in common with birds:—They are oviparous, or ovoviviparous, animals; the embryo is provided with the membranes termed amnion and allantois; there are no mammary glands; at no period of life are gills present; the skull is joined to the vertebral column by a single occipital condyle; each half of the lower jaw consists of several pieces; the lower jaw is not united directly with the skull, but to an intervening quadrate bone; the alimentary canal terminates in a cloaca; the red corpuscles of the blood are oval in shape and contain a central solid particle or nucleus;

the cavities of the chest and abdomen are not separated by a diaphragm; the hemispheres of the brain are not united by a *corpus callosum*.

On the other hand, reptiles differ from birds in the following particulars:—The covering of reptiles consists of horny scales, and sometimes, also, of bony plates, never of feathers; the fore-limbs never take the form of wings; the tarsal and metatarsal bones of the hind limb are never anchylosed; the heart generally contains three cavities, and the blood is cold: the bronchi of the lungs never terminate in air sacs.

43. Skeleton.—The vertebral column varies much in the different groups. Some serpents have as many as three hundred vertebræ, united by a "ball and socket" joint. The tail is generally very long. The limbs, when pre-

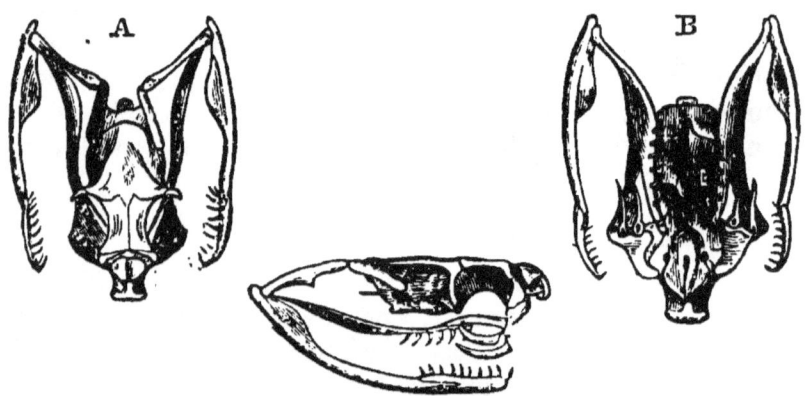

FIG. 14.—SKULL OF A VIPER.
a, view from above; *b*, from below; *c*, side view.

sent, are sometimes used in walking, sometimes in swimming, and in the extinct **pterodactyles** (Gr. *pteron*, a wing; *daktulos*, a finger) as organs of flight. In this last case, however, the skeleton of the fore-limb had no resemblance to the wing bones of a bird. The serpents and some lizards are destitute of limbs.

There is a single condyle on the occipital bone, and a corresponding facet on the atlas. There are from four to six pieces in each half of the lower jaw, which is united to the skull by a quadrate bone. The various pieces, as well as the two halves, are anchylosed in the tortoises. In the snakes the two halves of the lower jaw are only united by muscles and ligaments.

44. Digestion.—All reptiles, except the turtles and tortoises, possess *teeth*. The teeth in the various parts of the jaw are all of the same kind, and, except in the crocodiles, are not inserted in separate sockets. The tortoises and turtles have horny beaks instead of teeth. Generally speaking, there is not much peculiarity in the digestive system of the **Reptilia**. Salivary glands are sometimes absent. In serpents the stomach is but little different from the gullet and intestine; in crocodiles it bears a considerable resemblance to the gizzard of birds. The crocodiles also swallow pebbles to assist in grinding the food. The numerous conical teeth with which these animals are provided, are better adapted for holding the food than for masticating it. The intestine, as in birds, opens into a **cloaca**, which also receives the secretions of the kidneys and genital organs.

45. Circulation.—In mammals and birds the heart always contains four cavities—two auricles and two ventricles; and the right and left sides are always distinct, the chambers of the right side containing only venous blood, and the left side arterial blood. In reptiles the heart has always two auricles; sometimes one, and sometimes two ventricles. When there is only one ventricle, the following is the course of the circulation:—The impure venous blood from the body is poured by the **venæ cavæ** into the right auricle, thence into the ventricle; the arterial blood is conveyed by the pulmonary veins to the left auricle, and passes also into the ventricle. This cavity thus contains a mixture of venous and arterial blood. This mixed fluid is driven by the ventricle, partly to the lungs by the pulmonary artery, and partly by the aorta to all parts of the body.

In the crocodiles there is a complete partition between the two ventricles, the heart in this group containing four chambers. On this account, the mixture of arterial and venous blood cannot take place within the heart itself; but the pulmonary artery and aorta are connected by a small aperture near their origin. Through this aperture, the venous blood contained in the pulmonary artery is mixed with the arterial blood contained in the aorta. Thus, in the crocodiles, a mixed fluid is conveyed to the lungs and through the system, as well as in the other Reptilia.

On account of this peculiarity in the structure of the heart, and in the course of the circulation, the blood of the Reptilia, being imperfectly aerated, is cold; and, as a consequence, the movements of these animals are sluggish, and their intelligence of a lower order. "It may be remarked, however, that stupidity in reptiles is associated, as it often is in man, with a venomous and rancorous disposition; and that these defects, both intellectual and moral, seem to depend upon an imperfect oxidation of the blood corpuscles." (*Dr. Haughton.*)

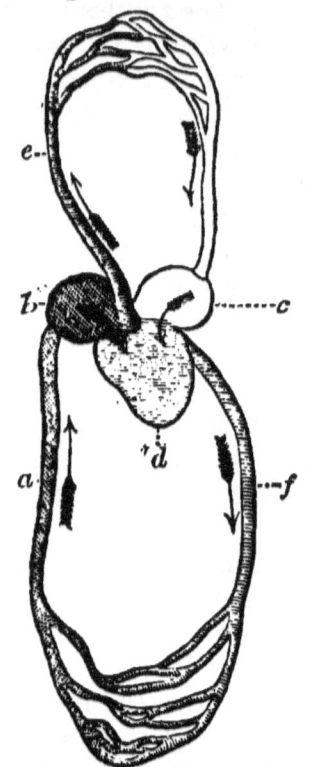

FIG. 15.—CIRCULATION IN REPTILES.

a, vena cava; *b*, right auricle, receiving venous blood from the body; *c*, left auricle, receiving arterial blood from the lungs; *d*, common ventricle, receiving mixed blood; *e*, pulmonary artery, conveying mixed blood to the lungs; *f*, aorta, conveying mixed blood through the system.

The red corpuscles of the blood are oval and nucleated.

46. Respiration.—The cavities of the chest and abdomen are not separated by a diaphragm. The lungs are large and often extend into the abdomen. Their bronchi are closed, and do not communicate with air sacs in the body. In the serpents, there is only a single lung.

47. Nervous System.—The brain is relatively smaller than in the birds or mammals. The surfaces of the hemispheres of the **cerebrum** are without convolutions. There is no **corpus callosum.**

48. Covering.—The covering of reptiles consists of horny scales developed by the **epidermis** (Gr. *epi*, upon ; *derma*, the skin) or cuticle ; and in the crocodiles, tortoises, and turtles, there are also bony plates, formed by the **derma** or true skin.

49. Development.—All reptiles are oviparous. The outer coating of the egg is usually leathery, instead of being calcareous as in the birds. Sometimes the egg undergoes the necessary incubation within the body of the mother, so that the young is developed before the exclusion of the egg. Such reptiles are said to be **ovoviviparous,**

Class IV.—Amphibia.

50. The term **Amphibia** (Gr. *amphi*, both; *bios*, life) was formerly applied to aquatic mammals, such as the seals; but it is now restricted to a group of animals, including the frogs, toads, and newts, which at one period of life are adapted for breathing in water, and afterwards for breathing in air.

The Amphibia and the fishes form the section of the Vertebrata termed **Ichthyopsida.** They resemble fish in the following details:—They are oviparous; the embryo is destitute of the membranes termed amnion and allantois; gills or organs adapted for breathing the air dissolved in water are always present at some period of life; there is no diaphragm; the red blood-corpuscles are oval and nucleated; there are no mammary glands.

51. Skeleton.—The **skull,** as in the mammalia, unites with the vertebral column by two occipital condyles; but the "basi-occipital" is cartilaginous—not bony, as in the higher Vertebrata. There are two corresponding cups in the atlas. This structure furnishes a ready means of distinguishing the skull of an amphibian from that of a reptile.

52. The **limbs**, when present, resemble in their structure those of the higher Vertebrata. In the fore-limb there is a **humerus,** followed by **radius** and **ulna, carpus, meta-carpus,** and **phalanges;** and in the hind limb a **femur, tibia,** and **fibula, tarsus, meta-tarsus,** and **phalanges.** These parts sometimes consist of bone; in other cases, of cartilage.

53. Digestion.—The mouth is of large size, and is generally furnished with small conical teeth. There is a liver, divided into two lobes; a gall-bladder, pancreas, and spleen. The intestine, which is short, opens, as in the Reptilia, into a **cloaca,** which also receives the secretions of the kidneys and generative organs.

54. Respiration.—The nature of the respiratory organs forms the most distinctive feature of the Amphibia. In the larval or young state, they all possess gills which enable them to breath the air contained in water. These gills are external filaments, situated on the sides of the neck. In the frogs, toads, and newts, there are two sets of gills; one external, which soon disappears, and one lodged in an internal chamber, which remains for a longer period. All the Amphibia possess lungs in the adult state. Generally the gills disappear when the lungs are developed, but sometimes the external gills are retained through life, the animal in this case possessing both lungs and gills.

The nasal sacs always open posteriorly into the mouth.

55. Circulation.—In the young state, so long as they breathe by gills, the circulation of the Amphibia is similar to that of fishes, the heart containing only two cavities, an auricle and a ventricle. When the lungs are developed, however, the auricle is divided, and the heart then contains three chambers, as in the greater number of reptiles. The venous blood from the body passes to the right auricle, and the arterial blood from the lungs to the left auricle; both these chambers pour their contents into the common ventricle, which drives this mixed fluid partly through the system, and partly to the lungs.

The blood is cold. The red corpuscles are oval, and contain a central nucleus.

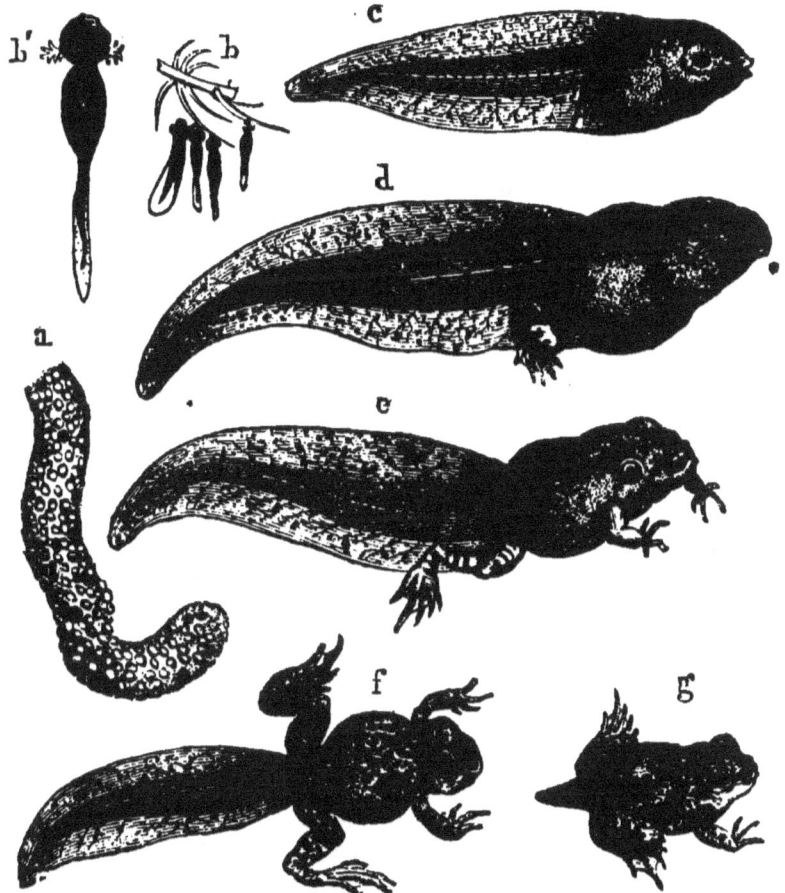

FIG. 16.—METAMORPHOSIS OF TOAD.

a, ova; *b*, tadpoles just hatched; *b'*, tadpole enlarged to show external gills; *c*, tadpole after losing external gills; *d*, the same after development of posterior limbs; *e*, provided with both pairs of limbs; *f*, tail begins to disappear; *g*, tail almost wholly absorbed.

56. Covering.—The covering consists of a soft moist skin, which is generally destitute of scales.

57. Development.—The Amphibia all undergo **metamorphosis** (Gr. *meta*, change; *morphe*, form). The eggs are deposited in water in a mass. They are destitute of shells. Each **ovum** consists of a black **yelk**, surrounded by a gelatinous "white." The ova are impregnated by the male immediately after exclusion. If a quantity of the spawn of the common frog is placed in water in a glass vessel, the following phenomena will be observed. After a time small fish-like animals will proceed from the dark specks of yelk. These, if viewed by a microscope, will be seen to be destitute of feet, to possess external gills in the form of branched tufts, and a long tail, surrounded by a "median fin," unsupported by rays. After a time the gills decay, the hind feet are developed, then the fore-feet. The tail is gradually absorbed into the body, and finally disappears altogether. In the newts the fore-feet make their appearance first, and the tail is retained.

Class V.—Pisces.

58. The **Pisces** or fishes (Lat. *piscis*, a fish) form the last class of vertebrate animals. These animals live entirely in water, and for this mode of life they are admirably adapted, the shape of their bodies, and the arrangement of the scales with which they are covered, presenting the least possible resistance to the medium through which they move.

Fishes form the second class of the section *Ichthyopsida*. They agree with the *Amphibia* in breathing by gills which are permanent; in being destitute of the membranes amnion and allantois in the embryonic state; in possessing cold blood, and in having oval, nucleated, red blood-corpuscles. They are also, like the Amphibia, oviparous animals.

They differ from the Amphibia in being covered with scales instead of having a naked skin; in possessing a heart with only a single auricle and a ventricle. In the fishes also, the nasal cavities do not open behind into the pharynx, as is the case with all the higher Vertebrata.

The gills remain throughout life, and the limbs take the form of fins.

59. Skeleton.—The internal skeleton of some fishes consists of cartilage or gristle; in others it is partly composed of cartilage and partly of bone. Most of the extinct fishes, especially of those found in the primary rocks, had cartilaginous skeletons; but the skeletons of the majority of modern fishes consist of true bone.

In cartilaginous fishes the various parts of the skull are joined together so as to form a single piece; but in the bony fishes the skull is very complex.

The number of vertebræ is very various, ranging in different fishes from seventeen to one hundred. In bony fishes, the vertebræ are **amphicœlus** (Gr. *amphi*, both; *koilos*, hollow) or concave at both ends. Their edges are united together by ligaments, and the hollow space between the vertebræ is filled with a gelatinous substance, so that the spine is extremely flexible. The vertebral column is divisible into the two regions—caudal and abdominal. The **ribs** are confined to the abdominal region. Their lower ends are free, the sternum being absent.

60. The **limbs**, when present, take the form of "fins," which are prolongations of the integument or skin, supported by bony or cartilaginous **rays**. The pair of fins which correspond to the arms of a man are called **pectoral** (Lat. *pectus*, the breast) fins. They are attached to bones which bear some resemblance to the fore-arm in the higher vertebrates. There is a **pectoral arch** in which the scapula, clavicle, and coracoid bones are represented. The humerus is wanting; but there are usually bones to represent the radius and ulna, and the carpus. To this latter the fin rays are attached.

61. The **ventral** (Lat. *venter*, the belly) fins correspond to the legs of a man. They are united directly to a kind of pelvic arch, the bones of the hind limb being unrepresented. Besides these fins, which are called **paired fins**, there are others which occupy the middle line of the body, and are, for this reason, termed **median fins**. These fins are joined to a series of pointed bones called **interspinous bones,** which are connected at their inner ends by ligaments with the spinous processes of the vertebræ. The median fins on the back, one or two in number, are called **dorsal** (Lat. *dorsum,* the back) fins;

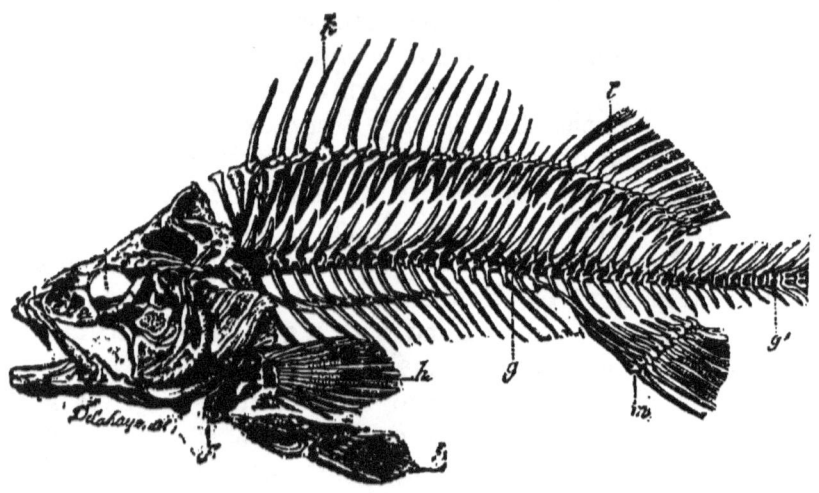

FIG. 17.—SKELETON OF THE PERCH.

f, operculum; *gg',* vertebral column; *h,* one of the pectoral fins; *i,* ventral fin; *kl,* first and second dorsal fins, supported on interspinous bones; *m* anal fin, with its interspinous bones.

those near the vent, **anal fins**; and the fin at the extremity of the vertebral column is termed the **caudal** (Lat. *cauda,* the tail) or tail fin. The median fins of fishes differ from those that are met with in the larvæ of the *Amphibia,* in being supported by "fin rays." The caudal fin, which is the principal agent in locomotion, is always placed vertically, and not horizontally as in the whale tribe.

There are two types of caudal fin, **homocercal** (Gr. *homos,* same; *kerkos,* a tail), and **heterocercal** (Gr. *heteros,* different; *kerkos,* a tail). In the homocercal type, the vertebral column terminates at the commencement of the tail lobes, which are of the same shape and of equal size. In the heterocercal type, the vertebral column extends to the extremity of the upper tail lobe. The greater portion of the caudal fin is in this case below the vertebral column, the lobes being unequal. Most modern fishes have homocercal tails, while the heterocercal type is commonest in fossil fishes. The sturgeons, sharks, &c., have heterocercal tails.

62. Digestion.—The teeth of fishes are not confined to the jaws, but are usually distributed over all the bones which assist in forming the cavity of the mouth They are not inserted in sockets, being merely attached to the surface of the bones. The gullet, which is short and wide, opens into a large stomach. The inner extremity of the stomach, where it opens into the intestine, is furnished with a valve, behind which are a number of blind tubes, termed the "pyloric cæca," which are thought to be the representatives of the pancreas. A true pancreas, however, is sometimes present. The intestinal canal varies in length in different fish. The liver is usually large, and contains much oil. The kidneys are also so large as to extend from the one extremity of the abdomen to the other.

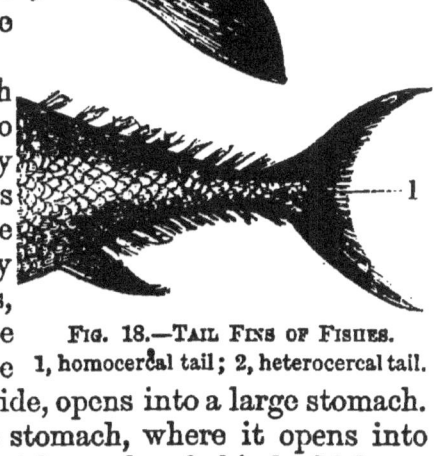

FIG. 18.—TAIL FINS OF FISHES.
1, homocercal tail; 2, heterocercal tail.

63. Circulation.—The heart generally consists of two cavities, an auricle and a ventricle. The exceptions are the Amphioxus or Lancelet, which has no heart, and the Lepidosiren or mud-fish, which has two auricles and one ventricle. The venous blood, after traversing the body, passes through the vena cava into the auricle which transmits it to the ventricle. It passes from the ventricle into the "branchial artery" which conveys it to the gills, where it is purified by the action of the air contained in the water. This artery, at its base, where it leaves the ventricle, usually swells out into a cavity furnished with muscular walls, termed the **bulbus arteriosus**. Its use is to assist the ventricle in propelling the blood to the gills. Instead of passing immediately to the heart, as in the other Vertebrata, the aerated blood is conveyed from the gills through the system, and is changed into venous blood before it again reaches the auricle.

The blood is generally of the same temperature as the water in which the fish lives. The majority of the corpuscles are red (except in the lancelet), oval, and nucleated.

64. **Respiration.**—All fishes are furnished with an apparatus termed gills, adapted for breathing the air contained in water. The following description of the structure and the arrangements of the gills applies only to the bony fishes. In the cartilaginous fishes the arrangements are considerably modified. The gills consist of cartilaginous leaflets arranged either in single or double rows, covered with a mucous membrane which is abundantly supplied with minute blood-vessels. These leaflets are supported on bony or cartilaginous arches which are separated from one another by slits, and are connected with the **hyoid** or tongue bone. The arches are situated in cavities on each side of the neck, termed the "branchial chambers." The fish takes a quantity of

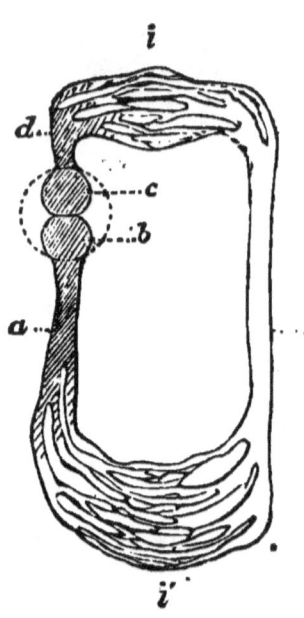

Fig. 19.—Circulation of a Fish.

a, system of veins, conveying the blood to the heart; *b*, auricle; *c*, ventricle; *d*, branchial artery, corresponding to the pulmonary artery of the higher vertebrates; *h*, aorta, conveying aerated blood through the system; *ii*, capillaries.

water into its mouth by a process allied to swallowing. This water is driven from the pharynx, through the slits, into the gill chambers, and having passed over the gills, escapes by apertures at each side of the neck, called the "gill slits." These gill slits are closed in front by the **operculum** (Lat. *operculum*, a lid) or gill cover. The blood contained in the mumerous vessels with which the gills are supplied, is aerated by the oxygen of the air contained in the water which thus passes over them.

Most fish are furnished with a sac filled with gas, called the "swim bladder." It lies between the alimentary canal and the kidneys. Its probable use is to enable the fish to maintain the proper specific gravity. In the mudfish it opens into the gullet by a tube, and seems to have some analogy to the lungs of the higher vertebrates.

65. Nervous System and Senses.—The **brain** of fishes is very small compared with the spinal cord. It consists of a series of lobes or ganglia connected by nerve fibres. There are one or two **nostrils**, to which water is admitted, but the nasal sacs are closed behind, and, except in the lepidosiren or mud-fish, do not communicate with the pharynx as in the higher Vertebrata. The organs of **hearing** consist of two cavities placed on the side of the head filled with fluid, and containing each two **otolithes** (Gr. *ous, otos,* an ear; *lithos,* a stone). There is no external ear.

66. Covering.—Fishes are covered with scales, which usually overlap one another, like the slates on the roof of a house. These scales are of four different kinds:—

1. **Cycloid** (Gr. *kuklos,* a circle; *eidos,* form).—These are thin, smooth, and horny, and are either circular or oval in shape. Most of our common fishes have cycloid scales.

2. **Ctenoid** (Gr. *kteis, ktenos,* a comb; *eidos,* form).— These are also thin and horny, but their hinder edges are fringed with comb-like spines. The scales of the perch are ctenoid.

3. **Ganoid** (Gr. *ganos,* splendour; *eidos,* form).—

Ganoid scales consist of a layer of bone covered with a coating of enamel. They are thicker and larger than the other kinds of scales. They are generally rectangular in form, and instead of overlapping, are in contact at the edges, like tiles in a floor. Few living fishes have ganoid scales, but they are characteristic of fossil fish, especially those found in the primary rocks. The sturgeon is a modern example.

4. **Placoid** (Gr. *plax*, a plate; *eidos*, form).—These scales consist of bony grains or plates scattered over the surface of the skin. The plates are sometimes furnished with acute spines. They are found in the sharks and rays.

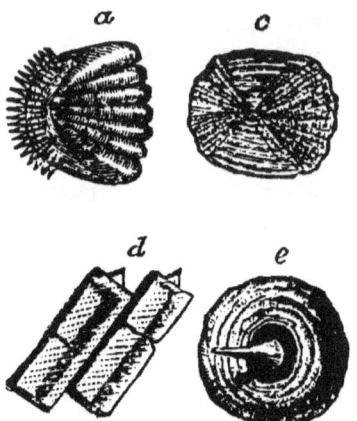

FIG. 20.—FISH SCALES.
c, cycloid scale of the carp; *a*, ctenoid scale of the perch; *d*, ganoid scales; *e*, placoid scale.

67. **Development.**—Fish are generally oviparous animals. The ovaries, which are known by the name of roe, are of large size, and sometimes contain an immense number of ova. The male organs are called "milt" or soft roe. The ova, after deposition in the water, are impregnated by the milt.

Some fish are ovoviviparous.

QUESTIONS.—II.

1. What is the derivation of *Mammalia*?
2. What are the most important characters of this class?
3. What are the principal regions of the vertebral column?
4. What is the general number of the cervical vertebræ, and what are the exceptions?
5. How many dorsal vertebræ are there?
6. How is the cavity of the chest formed?
7. What is the sternum?
8. Where are the lumbar and sacral vertebræ situated
9. What is meant by "anchylosed"?

10. What are the *caudal* vertebræ?
11. What mammals have less than two pairs of limbs?
12. Name, in their order, the various bones of the fore-limb.
13. How is the fore-limb joined to the trunk?
14. What are the *scapulæ?*
15. What are the clavicles?
16. Name, in their order, the bones of the hind-limb.
17. Describe the teeth of a typical mammal.
18. What mammals are destitute of teeth?
19. How many sets of teeth may a mammal have?
20. What is saliva, and how is it produced?
21. What is the gastric juice, and where is it formed?
22. What is chyme?
23. How is it changed into chyle?
24. What is the source of the blood?
25. Describe the heart and circulatory organs in the Mammalia.
26. What is the peculiarity of the red blood-corpuscles in this class?
27. Describe the process of respiration in mammals.
28. How do the lungs of mammals differ from those of birds?
29. What are *bronchi?*
30. What is the *pleura?*
31. Describe the mammalian brain.
32. What is the *corpus callosum?*
33. How are mammals covered?
34. What members of this group are without hair?
35. How is hair modified in the porcupines and hedgehogs?
36. Into what sub-classes have the Mammalia been divided?
37. What is the *placenta,* and what mammals are destitute of this structure?
38. What are the *Sauropsida?*
39. What characters are common to birds and reptiles?
40. How has the skeleton of birds been adapted for flight?
41. What are the distinctive features of the skull in this class?
42. Describe the cervical region of the spine.
43. The dorsal region.
44. Why are the dorsal vertebræ usually anchylosed?
45. What is the "plough-share" bone?
46. How did the *Archæopteryx* differ from ordinary birds?
47. What is the peculiarity of the sternum?
48. What is its form in birds which do not fly?
49. Describe the "pectoral arch."
50. What is the "merry-thought"?
51. What is the structure of the "fore-limb"?
52. What is the structure of the "hind-limb"?
53. Describe the "tarso-metatarsus."
54. How many toes have birds, and how are they arranged?

55. What birds have more, and what have less, than four toes?
56. What is the nature of the beak?
57. Describe the digestive apparatus in birds.
58. How does the gizzard differ in grain-eating and flesh-eating birds?
59. What is the *cloaca?*
60. Compare the temperature of birds with that of mammals.
61. How is the high temperature accounted for?
62. Describe the circulation in this class.
63. How does the heart of a bird differ from that of a mammal?
64. What are the peculiar features of the respiratory process in birds?
65. What are "pneumatic" bones?
66. What are the distinctive features of the brain in birds?
67. Give the peculiar characters of the eye and ear.
68. Describe a feather.
69. How do ostrich feathers differ from those of ordinary birds?
70. What is "incubation"?
71. How are the *Autophagi* distinguished from the *Heterophagi?*
72. Give examples of the class *Reptilia*.
73. What characters are common to reptiles and birds?
74. How do reptiles differ from birds?
75. What reptiles are destitute of limbs?
76. What reptiles have no teeth?
77. How do the teeth of reptiles differ from those of mammals?
78. How do the teeth of crocodiles differ from those of other reptiles?
79. Describe the structure of the heart and the circulation in reptiles.
80. How does the heart of a crocodile differ from the heart of other reptiles?
81. How does the venous blood become mixed with the arterial blood in the crocodile?
82. Describe the covering of reptiles.
83. What is meant by *ovoviviparous?*
84. Define the class *Amphibia*, and mention some examples of this class.
85. How do these animals resemble, and how do they differ from fishes?
86. Describe the skull, fore-limb, and hind-limb of an amphibian.
87. Describe the process of respiration in the larval and adult states respectively.
88. What is the structure of the heart, and the nature of the circulation in this class?
89. Describe the metamorphosis of the common frog.
90. How does a newt differ from a frog?
91. Define the class *Pisces*, and give examples.

QUESTIONS.

92. In what respects do fishes resemble, and how do they differ from the *Amphibia?*
93. What is the nature of the skeleton in fishes?
94. What is meant by "amphicœlus"?
95. What is the nature of the limbs in fishes?
96. What are "fins"?
97. How are the "median" fins distinguished from "paired" fins?
98. What pairs of fins correspond respectively to the legs and arms of a man?
99. By what names are the median fins which surround the body known?
100. What are "interspinous bones"?
101. What is the distinction between a "heterocercal" and a "homocercal" tail?
102. How does the tail of a fish differ from that of a whale?
103. Describe the digestive apparatus in fishes.
104. What are "pyloric cæca"?
105. Describe the heart, and the course of the circulation in fishes.
106. What is the "bulbus arteriosus"?
107. Describe the process of respiration in fishes.
108. How do gills differ from lungs?
109. What is the "operculum" of a fish?
110. What is the nature and use of the "swim bladder"?
111. Describe the brain of a fish.
112. How do the nostrils differ from those of other vertebrates?
113. Describe the organs of hearing in this class.
114. What is the nature of the covering of fishes?
115. What are the different forms of fish scales?
116. What is the "roe," and what is the "milt"?
117. How does the lancelet differ from ordinary fishes?
118. What are the peculiar features of the lepidosiren or mud-fish?

CHAPTER III.

ANNULOSA AND ANNULOIDA.

SUB-KINGDOM II.—**Annulosa.**

DIVISION I.—**Arthropoda.**

CLASS 1.—**Insecta.**

68. The highest class of **Arthropoda** is termed **Insecta** (Lat. *inseco*, I cut into). It is so called on account of the deep grooves which usually separate the bodies of these animals into three distinct parts.

69. **Definition.**—Insects are distinguished from the other classes of Arthropoda by the following peculiarities of structure:—There are three distinct sections in the body—head, thorax, and abdomen; the perfect animal has three pairs of legs attached to the thorax, which also bears one or two pairs of wings. The abdomen is always without limbs. There is one pair of antennæ. Respiration is performed by means of air tubes or tracheæ.

70. **Skeleton.**—As in the Annulosa generally, the hard parts of insects are external. The body is covered by an integument which is hardened by a horny substance, termed **chitine** (Gr. *chiton*, a coat). To this exoskeleton, which covers both body and limbs, the muscles are attached. There are never more than twenty ringed segments, or **somites** (Gr. *soma*, body), in the body. Five or six of these somites form the head, and three the thorax. There are usually eight or nine somites in the abdomen. The somites of the head are so completely amalgamated as to form a single piece. The segments of the thorax are also amalgamated, but not so as to be undistinguishable. The segments of the abdomen are movable upon one another.

The head bears a pair of jointed **antennæ** (Lat. *antenna*, the yard-arm of a ship). These have been considered by some naturalists organs of smell; by others, organs of touch; others again suppose them to be organs of hearing.

Each joint of the thorax bears a pair of limbs. There are from six to nine joints in each limb. The foot or tarsus has several joints. It is followed successively by the "shank," "thigh," and "hip" joints. The foot is usually terminated by two hooks.

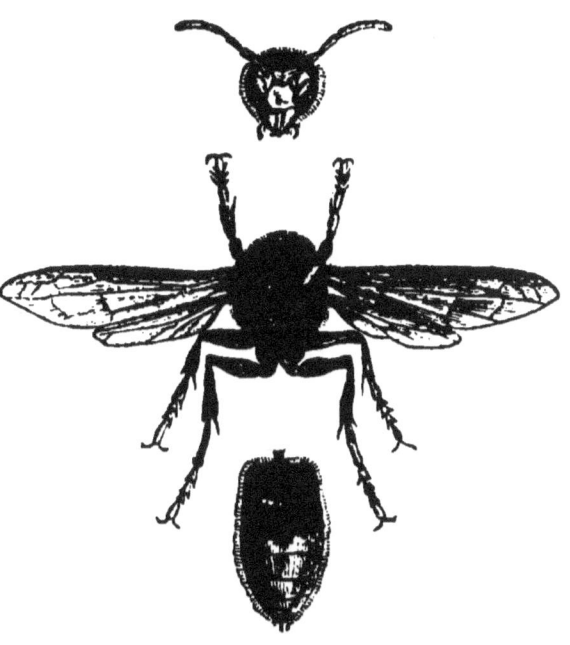

FIG 21.—DIAGRAM OF AN INSECT, With the division of its body into head, thorax, and abdomen.

There are generally two pairs of wings attached to the dorsal surfaces of the second and third segments of the thorax. The wings consist of "a highly elastic membrane, stretched over a frame-work of strong tubes, as the silk of an umbrella is expanded over its ribs." (*Gosse.*) The front pair of wings in the beetles are hardened by chitine, and form coverings for the protection of the membranous pair. The extremity of the abdomen is furnished with organs, which are sometimes used as weapons of offence and defence, as the sting in bees and wasps; and sometimes as "ovipositors," which serve for placing the eggs in the required position. This ovipositor is a very complex organ. It consists of a number of bristles enclosed in a sheath, the whole forming a tube through which the egg passes to its destination. In the working bees and wasps a similar tube forms the "sting," which not only makes a wound, but inserts poison therein.

Other insects are furnished with bristles at the extremity of the body. These, in the spring-tails, assist in leaping.

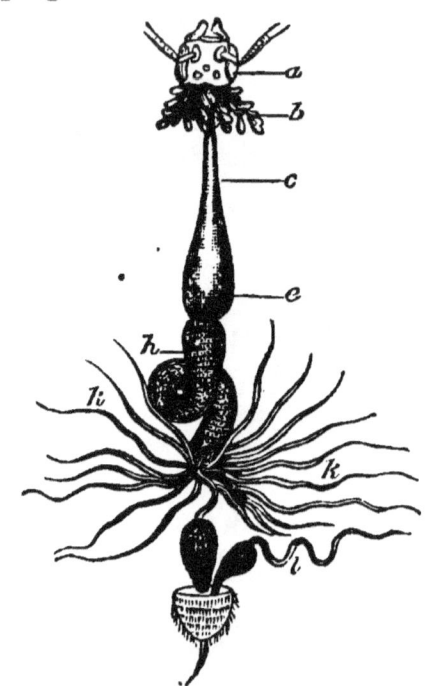

Fig. 22.—DIGESTIVE APPARATUS AND POISON GLAND OF THE BEE.

a, head and mouth; *b*, salivary glands: *c*, œsophagus; *e*, crop; *h*, stomach; *k*, Malphigian canals representing the liver; *l*, poison gland.

71. Digestion.— Some insects, such as the beetles, masticate their food; others, like the butterflies, subsist by suction. The organs of the mouth are varied to suit these different functions. In a beetle there are two lips, an upper (**labrum**) and a lower (**labium**), closing the mouth above and below; and two pairs of hooked and toothed jaws, which work horizontally. The upper jaws are termed **mandibles**, and are used in biting; the lower are called **maxillæ**, their office being to masticate the food.

The lower lip, which is believed to consist of a second pair of maxillæ united together, and the lower pair of jaws, bear each one or two jointed filaments termed **palpi** (Lat. *palpo*, I touch), which are supposed to be organs of touch.

In the butterflies, the upper lip and mandibles are undeveloped; the maxillæ are very long, and form, by their coalescence, a suctorial tube, which is coiled up under the head when not employed in sucking the juices of flowers. When coiled up, the tube is surrounded by the labial palpi, which, in these insects, take the form of two hairy cushions.

The mouth, in the bee, is adapted both for biting and suction. The labrum and mandibles are of the ordinary form, and are employed in the manufacture of the honeycomb. The maxillæ and the labium are converted into a long tube, which is used in sucking the honey which forms the food of these insects. This tube cannot be coiled up as in the butterflies.

In the bugs and plant lice the jaws are transformed into lancets. These are enclosed in the labium, which takes the form of a tubular sheath. In the flies (*Diptera*) the jaws sometimes take the form of bristles, and sometimes of lancets.

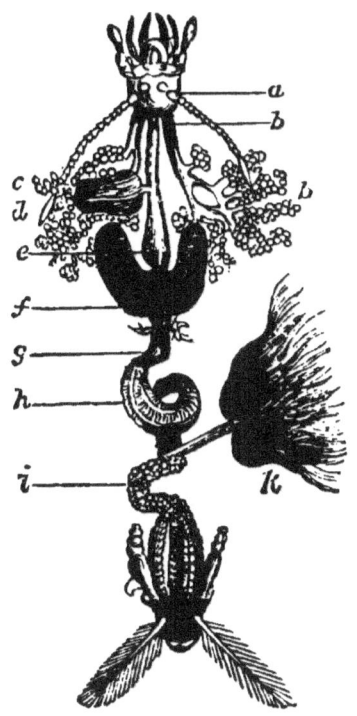

FIG. 23.—DIGESTIVE APPARATUS OF THE CRICKET.

a, head and its appendages; *bc*, salivary glands; *d*, antennæ; *e*, gizzard, preceded by the œsophagus bearing the crop; *f*, sacs connected with the gizzard; *gh*, true digestive stomach; *i*, intestine; *k*, Malphigian canals.

The mouth leads by an œsophagus into a membranous and generally folded stomach, termed the **crop**. From this, in the masticating insects, it passes into a second stomach, called the **gizzard**, which is furnished with muscular walls and plates of chitine, for grinding the food. Then follows the true digestive stomach, which is succeeded by an intestine of variable length, being short in suctorial insects, and long in those that feed on vegetables. The intestine ends in a **cloaca,** which also receives the products of the genital organs.

The œsophagus is usually furnished with tubular organs, which are believed to be **salivary glands.** There are certain other tubes connected with the intestine which are thought to be the representatives of the **liver** and **kidneys.**

72. Circulation.—The heart of insects is in the form of a long tube. It is placed along the back, and, on this account, is called the "dorsal vessel." It consists of eight or nine sacs opening into one another, and connected by valves which allow the blood to pass towards the head, but not in the opposite direction. The blood received into the heart is driven forwards by the contraction of these sacs, until it escapes near the head. As there are neither veins nor arteries, the blood is conveyed back to the heart through the interstices of the tissues.

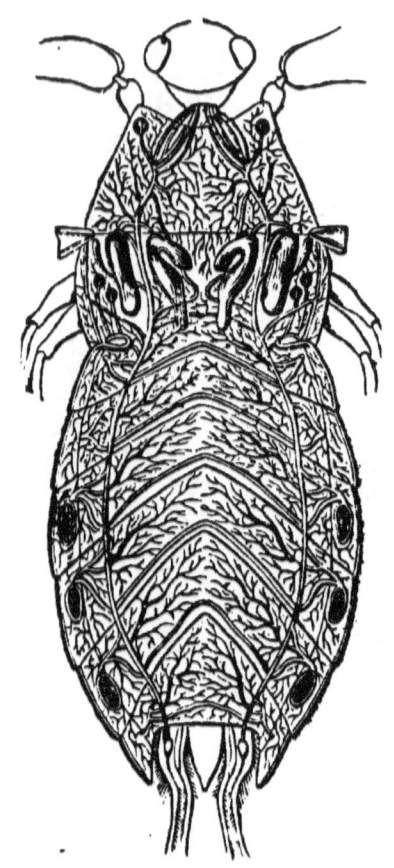

FIG. 24.—RESPIRATORY ORGANS OF AN INSECT.
Distribution of the tracheæ in the thorax and abdomen of the *Nepas*. Stigmata and air sacs.

73. Respiration is effected by a series of branched tubes called **tracheæ** (Gr. *tracheia*, the windpipe), which convey air through every part of the body. They communicate with the atmosphere by a number of slits, called **stigmata** (Gr. *stigma*, a mark), placed on both sides of the segments of the thorax and abdomen. There are seldom more than nine pairs of stigmata. The membranous walls of the tracheæ are kept distended by an elastic, chitinous thread which is coiled up within them in a spiral form. The blood in its passage through the tissues is purified by the oxygen of the air conveyed by the tracheæ.

74. **Nervous System and Senses.**—The nervous system in the Insecta bears a general resemblance to the form usually found in the Annulosa. It consists of a chain of ganglia, connected by a double nervous cord, occupying the ventral position of the body, and crossed in front by the gullet. The ganglia placed above the œsophagus, called the **cephalic** (Gr. *cephale*, the head) ganglia, or "brain," furnish nerves to the eyes and antennae. The ganglia below the œsophagus supply nerves to the legs and wings.

Fig. 25.—Spiral Thread of the Trachea of an Insect.

75. The **eyes** are usually compound, consisting often of many thousand hexagonal lenses. "They are formed by a large number of eye tubes closely pressed together, so that each of them has become hexagonal." (*Haughton.*) Each of these lenses is supplied with a nervous filament. There are usually simple eyes in addition to the compound eyes; and, in a few cases, these alone are present. These simple eyes consist of single hexagonal lenses. The only other organs of sense are the antennæ, which have already been described.

76. **Development.**—Insects are oviparous animals, and have the sexes distinct. The animal emerging from the egg has generally a very different form from that which it finally assumes. The changes which they undergo are termed **metamorphosis** (Gr. *meta*, change; *morphe*,

form). Some insects have a threefold metamorphosis; in others, the transformation is incomplete, and a few appear to remain unchanged. They have, accordingly, been divided into three sections, corresponding with the degree of transformation which they pass through:—

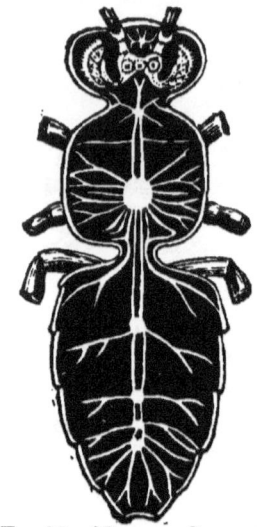

FIG. 26.—NERVOUS SYSTEM OF THE BEE.

1. **Holometabolic** (Gr. *holos*, whole; *metabole*, change) insects. These pass through three stages in progressing from the egg to the perfect insect. The first stage is termed the **larva** or caterpillar. In this state, the insect resembles a worm, and generally has numerous legs. It eats voraciously; and, as it increases in size, often changes its skin, acquiring a new covering of larger size as often as it does so. It then passes into a quiescent state in which it is termed a **pupa** (Lat. *pupa*, a doll) or chrysalis. Sometimes the pupa is covered with a "cocoon," spun by the larva; in other cases it is enclosed in a chitinous skin. While in this state, the legs, wings, and other organs are developed; and at length, the perfect insect, or **imago,** emerges from its prison. Beetles, butterflies, moths, &c., belong to this group.

2. **Hemimetabolic** (Gr. *hemi*, half; *metabole*, change) insects also undergo three stages of transformation. The larva, however, differs little from the imago, except in its being wingless, and of smaller size. The pupa is somewhat larger than the larva, and possesses rudimentary wings in the form of thick lobes, enclosed in cases. The pupa is active, instead of being quiescent, as in the holometabolic insects. Ultimately the wings are developed, and it then becomes the imago or perfect insect. This kind of metamorphosis is termed **incomplete.**

Grasshoppers, crickets, dragon-flies, &c., belong to this group. The larva and pupa of the dragon-fly live in water.

3. **Ametabolic** (Gr. *a*, without; *metabole*, change) insects undergo no metamorphosis. The adult insect does not differ from the young, except in point of size. These insects never acquire wings.

Class 2.—Myriapoda.

77. The **Myriapoda** (Gr. *murios*, ten thousand; *poda*, feet) is a very small class, comprising only the centipedes and millipedes. These animals in many respects resemble the insects, and are often included in that class. They differ from the Insecta, however, in the following characters:— The head is distinct, but there is no marked line of separation between the thorax and abdomen. There are always more than twenty somites in the body. The limbs are numerous, and are attached to the segments of the abdomen as well as to the thorax. The eyes are simple. Wings are never present.

78. **Skeleton.**—The myriapoda are provided with an integument hardened by chitine, as in the Insecta. This integument is divided into somites, of which there are always more than twenty. The head is always distinct, but there is no marked line of division between the thorax and abdomen. The head consists of five or six amalgamated somites. All the other segments of the body, except the last, are exactly like one another. The **limbs**, as the name of the class implies, are numerous, and are borne by the somites that correspond to the abdomen in the Insecta and Arachnida, as well as by those of the thorax. Each segment bears one pair of limbs. In one group, there are apparently two pairs of limbs attached to each segment. It seems, however, in these cases, that what appears to be a single segment, really consists of two somites amalgamated together. The appendages of the anterior segments are often modified into organs of prehension.

79. **The internal organs** of the myriapoda bear a close resemblance to those of the Insecta. There is a **dorsal**

vessel divided into sacs by valvular partitions, conveying the blood towards the head. Air is conveyed through

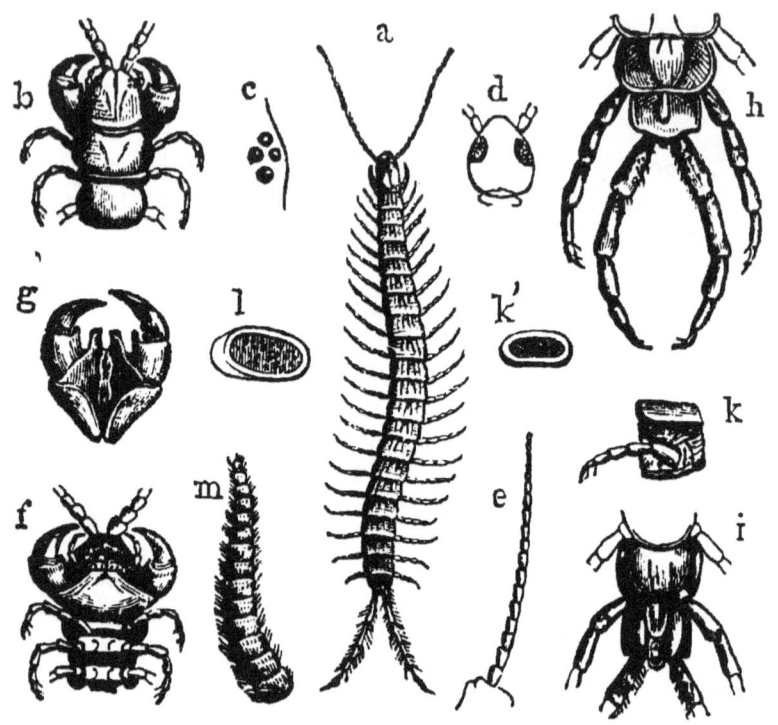

FIG. 27.—CENTIPEDE.

a, centipede; *b*, anterior part of the body seen from above; *c*, eyes of the right side magnified; *d*, head; *e*, antenna; *f*, anterior part of *scolopendra*, seen from below; *g*, perforated hooked jaws; *h*, posterior part of the body; *i*, the same seen from below; *k*, one of the somites with foot and stigma; *k' l*, isolated stigmata; *m*, antenna of *Geophilus*.

the body by tracheæ which communicate with the atmosphere by stigmata. In those groups which have one pair of limbs attached to each segment, the stigmata occur on the sides of every alternate somite; but when the segments bear two pairs of limbs, each is furnished with a pair of breathing apertures. This seems to corroborate the opinion, that in these cases, each segment consists of two amalgamated somites.

80. The **nervous system** is similar to that found in

the Insecta. There is one pair of antennæ attached to the head. The eyes are simple, and vary in number.

81. Development.—The young animal, on emerging from the egg, differs little from the adult, except that it has a smaller number of segments and limbs. At each change of skin it increases in size. Some of them, at first, are quite destitute of feet; others are furnished with three pairs, like the insects.

Class 3.—Arachnida.

82. The class **Arachnida** (Gr. *arachne*, a spider) comprises the spiders, scorpions, and mites. The head and thorax are united, forming a mass termed the **cephalothorax.** The abdomen is distinct. The antennæ are converted into a pair of mandibles or prehensile jaws. There are four pairs of limbs, borne by the cephalothorax. The abdomen is destitute of legs. Respiration is effected either by tracheæ or pulmonary sacs.

83. Skeleton.—The integument is usually hardened by chitine, but is sometimes soft and elastic. In the higher groups, the body is composed of twenty somites, six of which form the head. The segments of the thorax are amalgamated with those of the head, forming what is termed the **cephalothorax.** There are four pairs of limbs attached to the cephalothorax. The joints of the limbs are similar to those of the insects. Each limb is furnished with two claws at its extremity. The abdomen in the spiders is unsegmented; but in the scorpions it consists of a long, flexible, jointed tail. It is always destitute of legs.

The abdomen in the scorpions is terminated by a hooked claw, which is its principal weapon. This claw is perforated, and communicates with a poison gland situated at its base.

There are glands, situated in the abdomen of the spiders, which secrete a glutinous substance that has the property of hardening when exposed to the air. These glands communicate with four or six small teat-like organs, called **spinnerets**, placed at the extremity of the body, each of which contains a great number of minute tubes. Through each of these a slender thread of this sticky substance is drawn. The animal, by means of the claws of its hind feet, unites all these filaments together. The spider's thread is thus, small as it is, a rope composed of thousands of strands.

84. Digestion.—The mouth, which is situated in the front of the cephalothorax, is furnished with a pair of **mandibles**, which are believed to be the representatives of the antennæ of the other Arthropoda, a pair of **maxillæ**, and a **labium** or lower lip. In the scorpions there is also a **labrum** or upper lip. In the spiders, the mandibles are each provided with a perforated hooked joint, which conveys poison from a gland at its base. By means of this poison, the spiders are able to kill the insects, on the juices of which they subsist. The mandibles of the scorpions are terminated by pincers. The maxillæ are each provided with long jointed appendages, termed **palpi**. In the female spiders, these are terminated by claws; in the males, by sexual organs. The maxillary palpi of the scorpions are very long, and are furnished with grasping claws. As the food of the spiders consists merely of the juices of those animals on which they subsist, their alimentary apparatus is extremely simple. The intestine is short, and without convolutions. There are salivary glands, and tubes which are believed to represent kidneys.

85. Circulation.—There is usually a heart, similar to the "dorsal vessel" of the Insecta and Myriapoda. Some of the lower groups are without any special blood-vessels.

86. Respiration is performed variously in the different groups. All, however, breathe air directly. Some groups have tracheæ, like those found in the Insecta and Myriapoda; others have a modified form of these, consisting of a series of bags along the sides of the animal, formed by an involution of the integument, and communicating with the atmosphere by stigmata. The walls of these bags, which are termed **pulmonary sacs**, are abundantly supplied with blood-vessels. Sometimes there are both tracheæ and pulmonary sacs. In the lowest groups there are no special breathing organs, the blood being aerated by the general surface of the body.

87. The Nervous System consists of the usual chain of ganglia, but is more condensed than in the Insecta. There is a **cephalic ganglion** above the œsophagus, and a large ganglion placed in the thorax. There is sometimes also a small ganglion in the abdomen. These ganglia are united by a pair of filaments, and give off nerves to the various organs of the body.

88. The eyes, which are always simple, are placed in the front of the cephalothorax. They vary in number from two to eight.

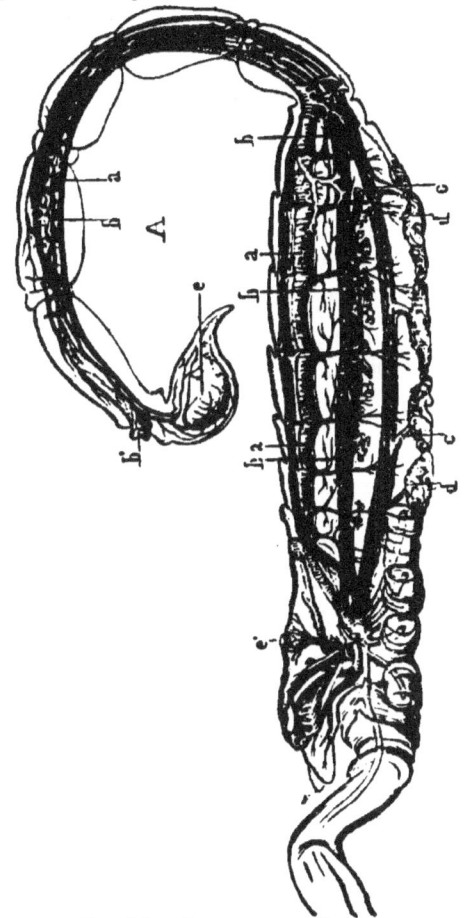

FIG. 28.—ANATOMY OF SCORPION.

a, the dorsal vessel and arteries; *b*, alimentary canal; *b'*, anus; *c*, chain of nervous ganglia; *e'*, one of the eyes and its optic nerve; *dd*, pulmonary sacs; *e*, terminal hooked claw with poison gland.

89. Development.—The Arachnida are generally oviparous, and, with some exceptions, unisexual. In most cases, the young animal resembles its parent in everything except size. The mites, however, have but three pairs of legs in the larval state, and only obtain the additional pair after casting the skin several times.

90. Divisions.—The Arachnida may be divided into two sub-classes:—

1. **Pulmonaria.**—These breathe by pulmonary sacs, and sometimes by tracheæ also. They have six or eight simple eyes. The scorpions and most spiders belong to this sub-class.

2. **Trachearia.**—The members of this group breathe either by tracheæ or by the general surface of the body. They have never more than four simple eyes. This sub-class includes the cheese-mite, itch-insect, and harvest-spider.

Class 4.—Crustacea.

91. The class **Crustacea** (Lat. *crusta*, a crust) is so called, because the body is often covered by a skin which contains a considerable quantity of carbonate of lime, and is hardened into a shell or crust. This skin forms the external skeleton, and within it all the soft parts of the body are inclosed. The Crustacea is a very large class, including lobsters, crabs, shrimps, prawns, barnacles, sea acorns, &c., groups of animals that differ widely in appearance and structure. It is closely allied to the Arachnida; but may be distinguished from it and the other classes of Arthropoda by the following characters:—When breathing organs are present, they always take the form of gills or branchiæ, which adapt them for respiring air in water; they have always *more than* four pairs of limbs (the usual number being five or seven pairs), which are not confined to the thorax, but are usually attached to the abdomen also; one or more of these pairs of limbs are usually modified into masticating jaws; there are two pairs of antennæ; they undergo metamorphosis before arriving at maturity.

92. Skeleton.—The body is usually divided into three

regions—head, thorax, and abdomen. In a typical specimen, it is composed of twenty-one somites. Seven of these are assigned to the head, thorax, and abdomen, respectively. The segments of the head and thorax are usually united into a single mass, termed the cephalothorax. The skeleton is so varied, that no general description can be given, which will apply to all the groups. The **lobster** may be selected as a representative of the class. In this animal the body is divided into two regions, which are popularly termed the **head** and **tail**. The so-called "head" is the cephalothorax, which is covered by a large plate termed the **carapace**. The "tail" is the abdomen, the joints of which are movable upon one another. To the first segment of the head are attached movable stalks which bear a pair of compound eyes. The next two segments bear two pairs of antennæ, called respectively the **antennules**, and the **great antennæ**. The appendages of the mouth follow, and are of a very complex character. There is a **labrum** or upper lip, a **labium** or lower lip, a pair of biting jaws or **mandibles**, and two pairs of **maxillæ**. One of these is furnished with a spoon-like organ which is used in causing a current of water to flow through the gill chamber. The maxillæ are followed by three pairs of **maxillipedes**, or **foot-jaws**, one of which is attached to the last segment of the head, the others to the first two segments of the thorax. The maxillipedes are modified limbs, used as masticating organs. The remaining five segments of the thorax bear five pairs of walking legs. The first three pairs are furnished at their extremities with nipping claws. The front pair are much larger than the others, and are termed the **great claws**. The two remaining pairs of legs terminate in a single point, being without pincers.

To each of the segments of the abdomen, except the last, paddle-like organs, called **swimmerets**, are attached. The segment at the extremity of the body is termed the **telson** (Gr. *telson*, a boundary), and is destitute of appendages.

93. Digestion.—The alimentary canal is usually without convolutions, running straight through the body, from the mouth to the anus. The œsophagus leads to a large stomach, which in some cases is furnished with a calcareous apparatus for grinding the food. There is generally a large liver, but no salivary glands have been detected.

94. Circulation.—The heart, when present, is placed above the alimentary canal. In the higher groups, it consists of a single cavity, which, by its contractions, drives the blood through arteries to the various organs, and to the gills, where it is purified by the air contained in the water. There being no capillaries, the blood, after leaving the arteries, is conveyed through the interstices between the organs of the body to the veins which carry it back to the heart. It follows, therefore, that the heart in this class contains arterial and not venous blood, as in the case of fishes. The heart in some of the lower forms resembles the "dorsal vessel" of the insects.

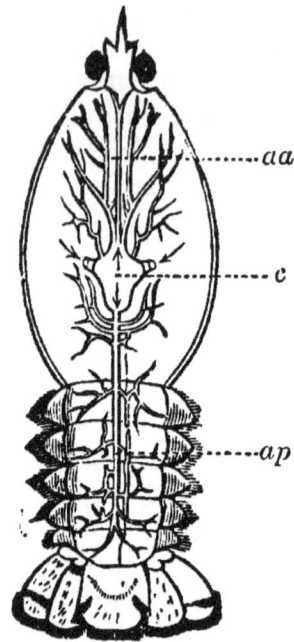

FIG. 29.—CIRCULATION IN THE CRAY FISH.

aa, anterior artery and its principal divisions; *c*, heart; *ap*, posterior aorta.

95. Respiration.—When the **Crustacea** possess distinct breathing organs, they always take the form of gills, adapted for obtaining the vital oxygen, not directly from the atmosphere, but from the air contained in water. In the higher groups, these gills are shaped like pointed pyramids, which consist of a central stem, bearing a great number of thin membranous plates placed closely together, but yet admitting a flow of water between them. These organs are abundantly supplied with blood-vessels. In the lobsters they are attached to the bases of the legs, and enclosed in a cavity formed beneath the **carapace** on each side of the body. These cavities are furnished with two openings, and circulation of water is kept up within them by the movements of the legs. In some of the lower groups, there are no distinct respiratory organs.

96. The **nervous system** is similar to that found in the Insecta. There is a pair of ganglia developed in each somite. These are connected by double nervous cords. The cephalic ganglia are placed above the œsophagus, the others run along the ventral surface of the body, the gullet passing between the two front pairs.

The higher Crustacea possess compound **eyes**, similar to those of the insects. They are often placed at the summit of movable stalks, which are sometimes of considerable length. In the lower forms, the eyes are simple. There is often only one eye, which is placed in the middle of the head.

97. The organs of **hearing** in the higher groups are placed close to the base of the long antennæ. They consist each of a hollow sac which is closed externally by a thin membrane. Behind this is a cavity filled with fluid, which is connected with the auditory nerve.

98. **Development.**—The Crustacea are oviparous animals, and, except one group, are all unisexual. The eggs, after exclusion, are generally carried about by the mother until near the period of hatching. They are either attached to the abdominal feet, or placed in pouches on each side of the tail. The young animals are often so unlike the adult, that, until the connection was observed, they were believed to be of different species. The early stages of the common shore-crab are thus described by Gosse:—" A hemispherical carapace or shell, not so big as a small pin's head, sends up from its centre a long pointed, curved spine, while another spine curves downwards from the front beneath the body like a beak ; the eyes are without stalks; there are two pairs of jointed

feet, ending in tufts of stiff bristles; and a long jointed body carried straight behind, which ends in two bundles of diverging spines. Such is the grotesque character under which our little masquerader makes his first appearance on any stage. After a time he drops his outer garments, and assumes a second form, widely different from the former, and still sufficiently remote from the ultimate one; and it is not till the third moult that the little creature, now grown to the size of a hemp-seed, begins to be recognizable as a crab; though even now he has several stages to pass through—several doffings of coats and trousers—before he is quite a proper shore-crab *comme il faut*."

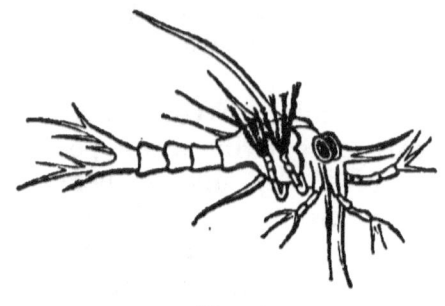

Fig. 30.
LARVA OF CRAB MAGNIFIED.

As these animals increase in size, the skin becomes too small to contain the body, and is cast off at intervals. Moulting in the insects is confined to the larval state, because when they reach the last stage, growth ceases; but the Crustacea continue to grow long after the perfect form is attained. While moulting, they retire to some secluded place, as if conscious of their helpless condition.

Division II.—Anarthropoda.
Class 1.—Annelida.

99. The **Annelida** (Lat. *annulus*, a ring) is a large class, containing the leech, the earth-worm, the lob-worm, and those animals which live in tubes, termed **Serpulæ**. It is distinguished from the Arthropoda mainly by the absence of jointed limbs.

The Annelida are worm-like animals, covered with a soft skin. The body is divided into a series of rings or segments, which are all exactly like one another, except the first and last. These rings are often very numerous, amounting in some groups to several hundreds. A ring or segment, in the higher groups, consists of two arches, a dorsal and ventral; and each of these bears two protuberances, placed on the sides of the animal, which have been called "foot tubercles." These tubercles support

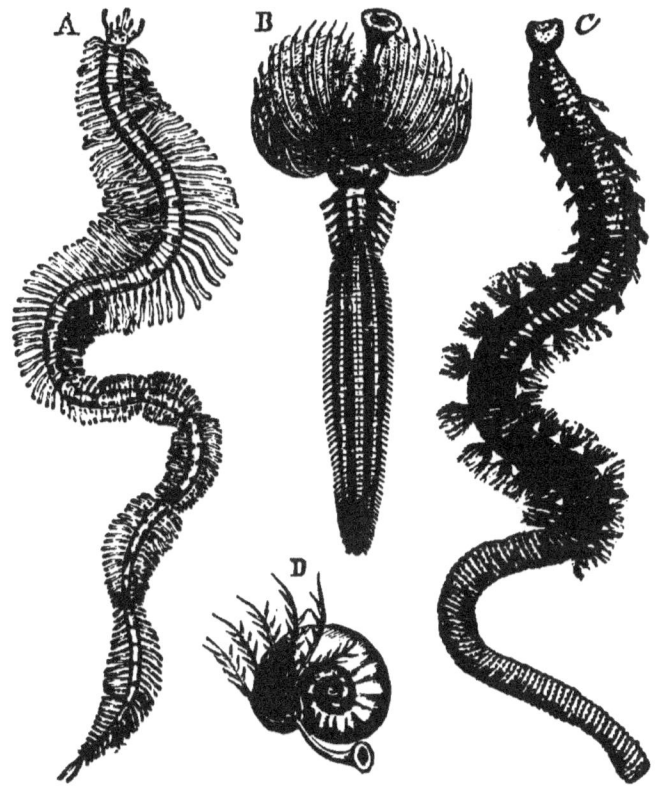

FIG. 31.—GROUP OF ANNELIDES.
A, Myrianida; *B*, Serpula; *C*, Lob-worm; *D*, Spirorbis.

bundles of bristles which surround a soft jointed filament, called the **cirrhus**. These bristles, by pressing against the ground, assist in locomotion. Jointed limbs are never met with in any animal of this class, nor are the organs of locomotion ever transformed into foot-jaws.

100. Digestion.—There is a mouth, sometimes furnished with horny jaws, a gullet, stomach, and intestine, which generally runs straight through the body, without convolutions, until it reaches the anal opening.

101. Circulation.—"No annelide ever possesses a heart comparable to the heart of a crustacean or insect; but a system of vessels, with more or less extensively contractile walls, containing a clear fluid, red or green in colour, and, in some cases only, corpusculated, is very generally developed, and sends prolongations into the respiratory organs, when such exist" (*Huxley*).

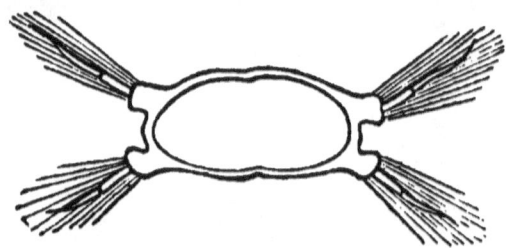

Fig. 32.—SECTION OF AN ANNELIDE.

This circulation is termed **pseudo-hæmal** (Gr. *pseudos*, false; *haima*, blood); and the vessels that contain this red or green fluid are considered extreme modifications of the "water vessels" of the **Annuloida**. There is also a corpusculated fluid contained in the spaces surrounding the internal organs (the **perivisceral** cavity), which is believed to correspond with the true blood system of the Arthropoda.

102. Respiration.—The breathing organs are either external gills of various forms, or bladder-like sacs placed on the sides of the body. "The external gills are well seen in the common lug-worm, where they consist of eleven pairs of arborescent organs, placed externally, and rather in advance of the middle line of the body." (*Haughton.*)

In some cases there are no special breathing organs, respiration being effected by the surface of the body.

103. The nervous system is of the usual annulose type. It consists of a double chain of ganglia running along the ventral part of the body, the gullet passing between the two anterior pairs. The front pair are termed the **cerebral ganglia.**

Simple **eyes** are possessed by most. They appear as small dark spots, generally on the front of the head, but sometimes on the gills. In other cases they are distributed over the various segments of the body.

There are usually two or more antennæ which are without joints.

104. Development.—The sexes are sometimes distinct, but the leeches and earth-worms are **hermaphrodite** (Gr. *Hermes*, Mercury; *Aphrodite*, Venus), that is, the sexes are united in the same individual. The embryos are generally furnished with vibratile cilia.

105. Divisions.—This class is generally divided into two groups :—

1. The **Branchiate** section, which includes the sand-worms and the serpulæ.

2. The **Abranchiate** group, comprising the earth-worms and leeches.

Class 2.—Chætognatha.

106. The class **Chætognatha** (Gr. *chaite*, hair; *gnathus*, a jaw) consists of only a single genus, Sagitta (Lat. *sagitta*, an arrow). These animals were formerly placed among the Annelida, but, according to Professor Huxley, "they are so unlike them and every other group as to require a class for themselves." They are small marine animals, about an inch in length, and with a transparent skin. The following is an abridgment of the account of these singular creatures furnished by Huxley, (see *Classification of Animals*, p. 53) :—

"The head is furnished with six pairs of bristles. Two of these are long and claw-like, and lie at the sides of the mouth. The other four are short, and placed in front of the head. The hinder part of the body is fringed by a fin-like membrane (hence the name *Sagitta*), which seems to be an expansion of the cuticle. The intestine is a simple straight tube extending from the mouth to the anus."

"A single oval ganglion lies in the abdomen, and sends forwards and backwards two pairs of lateral cords. The anterior cords unite in front and above the mouth into a hexagonal ganglion. This gives off two branches, which dilate at their extremities into the spheroidal ganglia, on which the darkly pigmented imperfect eye rests."

Sub-Kingdom III.—Annuloida.
Class 1.—Scolecida.

107. The class Scolecida (Gr. *scolex*, a worm) includes the **intestinal-worms** *(entozoa)*, the microscopic **wheel-animalcules**, the **hair-worms**, &c. These groups differ much in external appearance, but there are some important points of structure common to them all. Most of them possess what is called the **water vascular system**, a series of vessels which communicate with the exterior by one or more openings on the surface of the body, their branches permeating the interior. These vessels are believed to perform the double functions of circulation and respiration.

No heart or true circulatory apparatus has been observed in any of them.

108. The nervous system consists of only one or two ganglia.

109. There are seven groups included in this class— the **Rotifera** (or wheel-animalcules), the **Turbellaria**, the **Trematoda** (or flukes), the **Tæniada** (or tape-worms), the **Nematoidea** (or thread-worms), the **Acantocephala**, and the **Gordiaceæ**.

110. Rotifera (Lat. *rota*, a wheel; *fero*, I carry).—The Rotifera are microscopic animals, commonly met with in fresh water ponds, and infusions of vegetable matter, which have been exposed to the sun for a few days. They are very small, seldom exceeding the $\frac{1}{30}$ of an inch in length. Their true position in the animal scale is still a matter of doubt. They were formerly classed with the Infusoria; but the possession of a nervous system and a complex digestive apparatus, entitle them to a much higher place in the animal kingdom. Some place them among the Annelida; other observers consider them allied to the Insecta and Crustacea. They have been placed among the Scolecida, on account of the "water vascular system" which they possess. Their nervous system is also very different from anything found in the Annulosa.

Fig. 33.—Rotifera.

The body is covered with a membranous transparent skin, through which the internal organs can be accurately observed. Most are free swimming; but some are attached to aquatic plants, &c. The anterior part of the body is furnished with one or two discs surrounded by cilia, which move so rapidly as to produce the optical illusion of rotating wheels. Early observers believed that they really did rotate. The movements of these cilia, by producing currents in the water, bring supplies of those still more minute **Infusoria** on which the creatures feed. In the free swimming species they also form organs of locomotion, acting like the screw of a steamer. The body is shaped like a spindle or barrel. The lower extremity is furnished with a tapering foot-like organ, with many joints, which are capable of being sheathed one within another like a telescope. The "foot" either terminates in a sucking-disc, or is furnished with a pair of "toes." In either case it is able to attach itself to some object when it desires to become stationary.

111. Digestion.—As already mentioned, these animals are provided with very effective food-bringers—the cilia, with which the anterior end of the body is furnished. The mouth leads into a pharynx, which is furnished with a series of hammer-like pieces, believed by Gosse to be the representatives of the mandibles and maxillæ of the Insecta. The food having been crushed by these hammer-like jaws, passes through a gullet into a large stomach, which is succeeded by an intestine, terminating in a cavity called the **cloaca**.

112. **Circulation.**—Nothing corresponding to the heart and circulatory organs of the **Arthropoda** has been discovered in the **Rotifera**. There is, however, a well developed **water vascular system**. It communicates with the exterior by an opening leading into a **cloaca**, which also receives the excretions from the intestine and generative organs. This outlet communicates with two complex tubes termed **respiratory tubes**, which pass along the sides of the animal, and terminate near the head. No true organs of respiration have been observed.

113. **The nervous system** consists of a single mass, situated in the head of the animal. This mass is of such a large comparative size, that Gosse asserts, it "can only be compared to the brain of the Vertebrata." From this ganglion a pair of nervous threads proceed backwards. Such a nervous system is very different from the double ganglionated chain found in the Annulosa.

A red eye of simple structure is situated upon the ganglion.

114. **Development.**—In all the Rotifera the sexes are distinct. The description given above applies only to the females. The males are much smaller, and are entirely without digestive or masticatory organs. They, however, possess the **water vascular system** in common with the females. They live but a short time, their sole business being to secrete sperm.

The females deposit ova of a large comparative size. The young do not undergo metamorphosis.

115. **Turbellaria** (Lat. *turbo*, I disturb).—These animals derive their name from the currents they produce in

ANNULOIDA—SCOLECIDA.

water, by the movements of the cilia with which their bodies are furnished. The best known group are the **planaria** (Gr. *planē*, wandering). They are small, flattened animals, with soft, black, gelatinous flesh. They are found in great abundance in rivers and ponds, attached to submerged plants. Other larger species belong to the sea shore. The mouth is placed near the centre of the body. The digestive canal has only one opening. The skin is furnished with numerous cilia. There are two nervous ganglia placed near the mouth, connected by a filament. They possess a number of rudimentary eyes, which are situated on the front part of the body. There is a water vascular system which communicates with the exterior.

The **nemertés** or ribbon-worms also belong to this group. They are found on the sea shore coiled up in knots under stones, and are sometimes ninety feet long. The alimentary canal has two openings. The **water vascular system** in the adult seems to have no communication with the exterior.

116. Trematoda (Gr. *trema*, an opening) or flukes are internal parasites inhabiting the bodies of man, sheep, birds, and fishes. They are somewhat flat in shape, and are furnished with one or more suckers, by which they attach themselves to the bodies of their victims. One of these suckers leads into the mouth, and is succeeded by a branched alimentary canal, which has no anal opening. There is also a branched **water vascular system,** which communicates with the exterior. The best known of these animals infest the gall bladder or the liver of the sheep, producing the disease called **rot.**

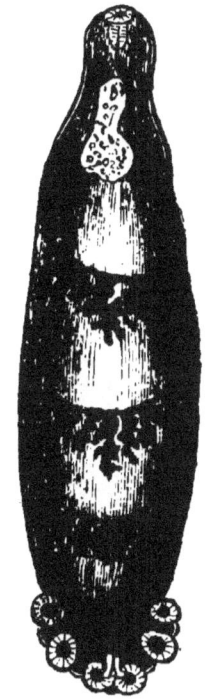

FIG. 34. TREMATODA.

82   ZOOLOGY.

117. Tæniada (Gr. *tainia*, a ribbon) or tape-worms are long ribbon-shaped worms, inhabiting the intestines of man and other warm-blooded animals. There are three different species found in the human intestines. The most common kind has been called **tænia solium**, because it was erroneously supposed that only one such worm was ever found in the intestine at the same time. The head of the tape-worm is furnished with suckers and recurved hooks, by which it attaches itself to the intestine of its host. This head, which is in reality the true animal, has no mouth nor digestive organs, its nourish-

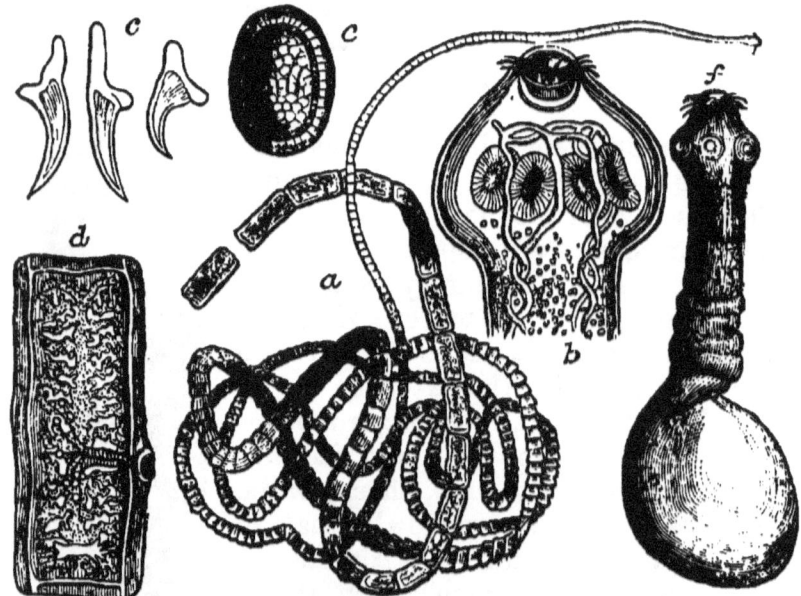

FIG. 35.—TÆNIA OR TAPE-WORM.

a, Tænia or tape-worm; *b*, head (enlarged) with hooks and suckers; *c*, hooks magnified; *d*, joint, showing branched ovary, generative pore, and water vascular canals; *e*, ovum; *f*, cystic worm (*cysticercus longicollis*).

ment being obtained by absorption. It is also destitute of reproductive organs. **The nervous system** is said to consist of one or two ganglia; but there are doubts on this point. There is a **water vascular system**, which consists of two tubes placed along the sides of the body. These vessels are connected by cross tubes at each joint.

The whole system communicates with the exterior by an aperture placed at the extremity of the last joint.

The head, by a species of budding, develops a series of flattened joints, each of which contains male and female organs. The joints nearest the head are always the newest, the mature joints being situated at the other extremity. The compound animal thus formed is often several yards long.

Each segment contains a large number of ova; but these are never hatched within the animal inhabited by the tape-worm. The mature joints gradually break off, and pass from the intestine; and the minute ova contained therein are swallowed by some warm blooded animal. In the case of the human tape-worm, the ova find their way into the stomach of the pig, where they are hatched. The embryo, by means of the flinty hooks with which its head is furnished, makes its way through the walls of the stomach, into the muscles. In this situation it develops at its lower extremity a bladder-like sac, filled with fluid. When it abounds in the flesh of the pig, the pork is said to be **measled**. These "measles" were formerly called **cystic** (Gr. *kustis*, a bladder) worms, and were believed to be different animals. These cystic-worms, so long as they remain in the tissues of the pig, never breed tape-worms, nor develop ova, but they are said to be able to multiply their own kind by budding. When a quantity of measly pork is eaten, a cystic-worm is sometimes transferred to the human stomach. It fastens itself by its hooks and suckers to the intestine, loses its bladder, and commences to develop joints by budding, and finally produces a tape-worm.

Another species of human tape-worm is derived from "measles" found in the muscles of the cow. The cystic-worms found in the tissues of the mouse and rat produce tape-worms in the intestines of the cat; and the cystic-worm which causes the disease called **staggers** in sheep, when transferred to the stomach of the dog, is developed into a tape-worm.

118. Nematoidea (Gr. *nema*, thread; *eidos*, form).—To this group belong the **round-worms, thread-worms, vinegar eels,** &c. Most of them infest the intestines of warm blooded quadrupeds. The round-worm sometimes attains a length of fifteen inches in the human intestine. The thread-worm inhabits the large intestine. It is found principally in children. The **Trichina** passes its early stage in the muscles of the pig, and is often transferred to the human stomach where it is developed. Other species infest the intestines of the horse and pig. These animals possess a mouth and an alimentary canal with anal opening. They have also a water-vascular system. A ganglionated nervous cord surrounds the gullet, from which nervous threads proceed towards the extremity of the body. The sexes are distinct.

There are many species of non-parasitic **Nematode** worms found in fresh water or on the sea shore. The **vinegar eel** is a well known specimen.

119. Acanthocephala (Gr. *acanthos*, thorn; *cephale*, the head).—These are formidable internal parasites, which inhabit the alimentary canal of many mammals, birds, and fishes; but have not yet been found in the human intestine. They are so called on account of the recurved hooks with which their proboscis or snout is surrounded. By means of these they attach themselves to the mucous membrane of the intestine. There is a single nervous ganglion at the base of this proboscis. There is a network of canals beneath the skin, containing a clear fluid, which is believed to be a modified form of the water-vascular system. They have neither mouth nor alimentary canal. The sexes are distinct.

120. Gordiaceæ (so called from the "Gordian Knot," because they twist their bodies into knots), or hair-worms, are popularly believed to be living horse hairs. They pass the early stages of their lives in the interior of insects, especially grasshoppers and water-beetles. They attain a length of ten or eleven inches, and then leave the body of their victim to breed. They deposit their eggs in water, or in some moist place, in long chains. They have a mouth and alimentary canal, but no anus. The sexes are distinct.

Class 2.—Echinodermata.

121. The class **Echinodermata** (Gr. *echinos*, hedgehog; *derma*, skin) includes the sea-urchins, star-fishes, sand-stars, brittle-stars, sea-cucumbers, and the almost extinct encrinites or sea-lilies, the remains of which are abundantly met with in the older rocks.

In the mature state, the members of this class are characterized by a radiate arrangement of the various parts, although in the embryonic condition they have **bilateral symmetry.** The covering is composed of numerous calcareous plates, joined together at the edges; or of a leathery integument, in which there are calcareous grains or spines. There is a **water-vascular system,** connected with a series of tube feet, termed the **ambulacral** (Lat. *ambulacrum*, a walking place) system. In all cases the alimentary canal is completely shut off from the cavity of the body, and is usually furnished with an anal opening. A ganglionated ring of nervous matter surrounds the gullet, from which branches radiate to the different parts of the body.

122. Skeleton.—The form of these animals varies much in the different groups. The sea-urchins are sometimes almost globular, the star-fishes are usually stars of five rays, the sea-cucumbers (*holothuria*) are worm-like. The covering of the sea-urchins consists of a shell or **test** which is composed of a great number of five or six-sided calcareous plates, joined together at the edges. Of these plates there are ten double rows or zones. In five of these zones the plates are small, and are perforated, so as to allow the emission of the tube-feet. These are called the

ambulacral plates. In the other five zones which alternate with the ambulacral rows, the plates are larger, and are not perforated. The shell is increased in size by additions made to the edges of each individual plate. This is effected by a delicate membrane which covers the test, and dips in between the edges of the various plates, adding calcareous matter to each so evenly as to preserve the shape of the whole.

Besides the plates which occur in zones, or lunes, there are others at the base and apex of the test. The mouth is surrounded by a leathery membrane which contains a number of small calcareous pieces called the **oral plates**; at the other extremity are similar **anal plates**. At the apex of the test, and surrounding the anal opening, there is a disc composed of a series of plates, termed the **genital** and **ocular** plates. There are five genital plates, which are larger in size than the ordinary plates, and pentagonal in shape. Each of them is perforated by a duct from the genital organs. One of these genital plates, larger than the others, is furnished with a protuberance which contains a large number of very minute openings. Through

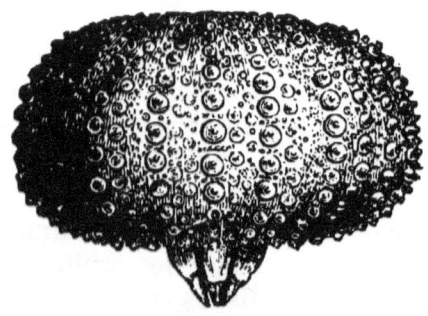

Fig. 36.—Test and Jaws of Sea-Urchin, with the Spines Removed.

these the **water-vascular system** communicates with the exterior. The apertures are so small, that the water in passing through them is freed from any particles of sand, &c., with which it may be mixed. This protuberance is

ANNULOIDA—ECHINODERMATA.

called the **madreporiform tubercle**, because, when examined by a lens, it somewhat resembles a piece of madrepore coral. The ocular plates, of which there are also five, are smaller than the genital, and are placed at the extremities of the ambulacral passages. Each plate contains an opening in which a little **eye** is placed.

The plates are furnished with numerous tubercles, to which pines are attached by a **ball-and-socket** joint.

FIG. 37.—PRINCIPAL ORGANS OF SEA-URCHIN.

a, œsophagus; *b e f*, intestine; *g*, anus; *h*, ovary; *i*, shell or test.

These spines assist in locomotion, and are also used as weapons of defence. Among the spines are found a number of curious appendages called **pedicellariæ** (Lat. *pedicellus*, a louse), because they were formerly believed to be parasites. They are long and worm-like, and at the head of each are placed three calcareous jaws, which fit together. These jaws continually open and shut. Their object seems to be to drive any parasites away which might attach themselves to the spines.

The covering in the star-fishes is leathery, but it contains calcareous plates, with tubercles and spines. The plates, however, do not unite at the edges, as in the sea-urchins.

123. Digestion.—The sea-urchins are vegetable feeders. They are furnished with five long calcareous teeth, meeting in a point. This structure has been termed **Aristotle's lantern.** The mouth, which is usually placed at the base of the shell, leads to a gullet, stomach, and intestine. This is much convoluted, winding round the interior of the shell. The anus is often, though not always, placed in the apex.

The star-fishes are animal feeders, and subsist usually on bivalve mollusks. The mouth is placed in the centre of the lower surface, and is destitute of teeth. The stomach is prolonged into the various rays. When there is an anal opening, it is situated on the upper surface.

124. Circulation.—In the sea-urchins the water-vascular system communicates with the exterior by the **madreporiform tubercle**, situated on the largest genital plate. From this tubercle proceeds a canal, which leads to a tube, forming a ring round the mouth, termed the **circular canal.** From this, five "radiating canals" proceed along the ambulacral areas. These canals give off short tubes, furnished with sucking discs, which are protruded through apertures in the shell. They are called **tube-feet**, because they assist in locomotion. At the base of the tube-feet are situated little bladders containing water, by which they are protruded or retracted at will.

Besides the water-vascular system, there is a true blood-vascular system. It consists of a central, contractile cavity or heart, which gives off vessels that form rings round the alimentary canal, near the mouth and anus respectively. From the anal ring, five arteries proceed along the ambulacral zones. These communicate with five vessels which run in the opposite direction, and convey the blood back to the heart. The details of the circulation in the star-fishes are similar in all essential features.

125. Respiration.—There are no distinct organs of respiration. The aeration of the blood is believed to be accomplished partly by the sea water with which the perivisceral cavity is filled, and partly by the **water-vascular system.**

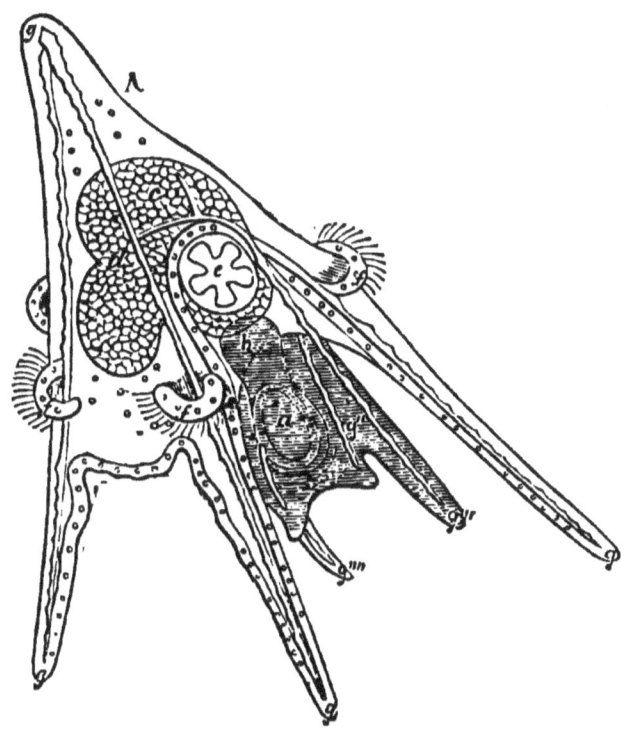

Fig. 38.—Larva of Sea-Urchin (Enlarged 160 times).

a, mouth ; *b*, stomach ; *c d*, intestine ; *e*, calcareous plate ; *f*, ciliated organs, assisting in swimming and respiration—perhaps the future ambulacral feet ; $gg'g''g'''$, calcareous stalks, serving as a frame-work for the body of the animal.

126. Nervous System.—The gullet is surrounded by a nervous cord, which connects a series of ganglia. From this, filaments proceed along the ambulacral areas.

127. Development.—The sexes are always distinct.

One of the most remarkable features of this class consists in the metamorphosis which the embryo undergoes. The following eloquent description of the metamorphosis of the brittle star is by Mr. Gosse:—

"The first condition of every echinoderm is the same —an egg-like body, covered with cilia, resembling an infusorium. Changes take place, and we presently see another form assumed, which varies in some degree in the different families. We lately had the pleasure of finding in our dip-net several little larvæ of a brittle star, the first that had ever been seen in our seas; and one of these we will select for description.

"A painter's long easel affords the only object with which to compare the little creature; for it consists of four long slender calcareous rods, arranged two in front and two behind, with connecting pieces going across in a peculiar manner, and meeting at the top in a slender head.

"On this shelly, fragile, and most delicate framework, as on a skeleton, are placed the soft parts of the animal, a clear gelatinous flesh, forming a sort of semi-oval tunic around it, from the summit to the middle, but thence downwards the rods individually are merely encased in the flesh without mutual connection. The interior of the body displays a large cavity, into which a sort of mouth ever and anon admits a gulp of water. Delicate cilia cover the whole integument, and are particularly large and strong on the flesh of the projecting rods.

"The appearance of this most singular animal is very beautiful, its colour pellucid white, except the summit of the apical knob, and the extremities of the greater rods, which are rose colour. It swims in an upright position, with a calm and deliberate progression. The specimens which we have seen were not more than one-fortieth of an inch in length.

"From this form the brittle star is developed, but in a manner unparalleled in any other class of animals. The exterior figure is not gradually changed, but the star is constructed within a particular part of the body of the larva, 'like a picture upon its canvas, or a piece of embroidery upon its frame.' The plane of the future starfish is not even the plane of the larva, but one quite independent of and oblique to it. Strange to tell, the young star does not absorb into itself the body of the larva, which has acted as a nidus for it, but throws it off as so much useless lumber—flesh, rods, and all."

It is a remarkable circumstance in the development of the Echinodermata that, while the adult forms are characterized by "radial symmetry," the young " exhibit as complete a 'bilateral symmetry' as annelids or insects."

QUESTIONS.—III.

1. What is the derivation of "Insecta"?
2. Give the definition of this class.
3. What is the nature of the skeleton?
4. What is *chitine?*
5. What is an *exoskeleton?*
6. What is a *somite?*
7. How many somites are there in the skeleton of an insect, and how many of these are assigned to each region of the body?
8. What are *antennæ*, and what is supposed to be their function in the Insecta?
9. What are the appendages of the thorax?
10. Describe the wings.
11. What is the peculiarity of the anterior pair of wings in the beetles?
12. What is the structure of the sting of a bee?
13. What are *ovipositors?*
14. Enumerate and describe the appendages of the mouth in a typical insect?
15. What are *palpi?*
16. How are these appendages modified in butterflies and bees?
17. How are the jaws modified in the bugs and flies?
18. Describe the digestive apparatus of a masticating insect.
19. Describe the circulatory organs in this class.
20. How does an insect breathe?
21. What are *tracheæ?*
22. What are *stigmata?*
23. Describe the nervous system of insects.
24. What are compound eyes?

25. How do *holometabolic* differ from *hemimetabolic* insects?
26. Describe the metamorphosis of each of these groups.
27. Explain the terms *larva, pupa,* and *imago.*
28. What is a cocoon?
29. Why are the *Myriapoda* so called?
30. Give the definition of this class.
31. Give examples.
32. In what respects do they differ from the *Insecta?*
33. What is the nature of the metamorphosis in the *Myriapoda?*
34. What animals are included in the class *Arachnida?*
35. Define the *Arachnida.*
36. What is the *cephalothorax?*
37. How do scorpions sting?
38. What are *spinnerets?*
39. Describe a spider's thread.
40. Describe the appendages of the head in the spiders and scorpion.
41. How do spiders kill their prey?
42. Describe the process of respiration in the *Arachnida.*
43. What are *pulmonary* sacs?
44. What is the nature of the nervous system?
45. Describe the organs of sight.
46. Describe the development of the *Arachnida.*
47. What are the principal divisions of this class, and how are they distinguished?
48. Define the class *Crustacea.*
49. Describe the exoskeleton of the lobster.
50. What is the "carapace"?
51. What is meant by the "head" and the "tail" of the lobster?
52. How many pairs of antennae have the *Crustacea?*
53. What is the usual number of limbs in this class?
54. What are swimmerets?
55. What are "foot jaws"?
56. What is the "telson"?
57. Describe the heart and circulatory organs of the lobster.
58. Describe the breathing organs of the lobster.
59. What is the nature of the eyes in the higher crustacea and in the lower groups?
60. Describe the organ of hearing.
61. Give an account of the metamorphosis of the common shore crab.
62. Define the *Annelida,* and give examples.
63. Describe the organs of locomotion possessed by some members of this class.
64. What is the nature of the digestive organs?
65. What is the nature of the circulatory fluid?
66. What is meant by the "pseudo-hæmal" system?

67. What is the nature of the respiration?
68. Describe the nervous system.
69. What sort of eyes do they possess?
70. What are the principal sub-divisions of this class?
71. Give an account of the class *Chætognatha*.
72. Define the class *Scolecida*.
73. What are the principal groups included in this class?
74. Give examples of each.
75. Why are the *Rotifera* so called?
76. Why have they been removed from the class *Infusoria*?
77. Why have they been placed in the class *Scolecida*?
78. In what respects do the males differ from the females?
79. Describe the digestive organs of the *Rotifera*.
80. What is the nature of the nervous system in this group?
81. Give an account of the *Turbellaria*.
82. Describe the *Planaria*.
83. Define the *Trematoda*, and give examples.
84. What produces the "rot" in sheep?
85. To what group does the "tape-worm" belong?
86. What are "cystic" worms, and what relation do they sustain to the tape-worms?
87. Describe the ordinary tape-worm.
88. Give an account of the development of this animal.
89. What are the "measles" met with in diseased pork?
90. Give examples of *Nematode* worms.
91. How do they differ in structure from the *Trematoda*?
92. Describe the *Acanthocephala*.
93. Why are they so called?
94. What are the *Gordiaceæ*?
95. Where do they spend their early life?
96. What are they commonly called?
97. Why is the class *Echinodermata* so called?
98. What animals are included in this class?
99. What is meant by "bilateral symmetry"?
100. Describe the "test" of a sea-urchin.
101. How does this test increase in size?
102. What are the "ambulacral" areas?
103. Where are the "genital" and "ocular" plates placed?
104. Why are they so called?
105. What is meant by the "madreporiform tubercle"?
106. Describe the "tube-feet."
107. What are "pedicellariæ," and what is their supposed function?
108. What is the nature and use of the spines?
109. How does the covering of the star-fishes differ from that of the sea-urchins?
110. Describe the digestive organs of the sea-urchins.

111. Describe the "water-vascular system."
112. What is meant by "Aristotle's Lantern?"
113. What is the course of the circulation in the sea-urchins?
114. What is the nature of the respiration?
115. Describe the nervous system.
116. Give an account of the metamorphosis of the "brittle star."

CHAPTER IV.

MOLLUSCA AND MOLLUSCOIDA.

SUB-KINGDOM IV.—Mollusca.

CLASS 1.—Cephalopoda.

128. The **Cephalopoda** (Gr. *kephale*, the head; *poda*, feet) is the highest class of **Mollusca**. It includes the cuttle-fish, the octopus or poulpe, the calamary or squid, the paper-nautilus, the pearly-nautilus, the extinct ammonites, belemnites, &c.

This class is characterized by having the organs of locomotion, in the form of arms, arranged in a circle around the mouth. These arms are usually furnished with sucking cups, and are eight or more in number. The body is covered with a loose mantle, which opens in front, on the lower part of the body, to admit water into the gill chamber. The gills are plume-like, and are either two or four in number. They are placed within the mantle. When the water in the gill chamber has been deprived of its oxygen, by the process of respiration, it is expelled through a "funnel" placed under the head. The reaction produced by the escape of this water causes the animal to move in the opposite direction.

The **Cephalopoda** are all carnivorous animals, and live in the sea. They walk along the sea bottom by means of their long arms; or swim, partly by the aid of a fin with which the extremity of many of their bodies is furnished, and partly by the reaction caused by the escape of water through the funnel.

129. **Skeleton.**—The greater number of living cephalopods are unprotected by any external shell. Most of these, however, possess an internal shell, but this is sometimes rudimentary.

96 ZOOLOGY.

In the cuttle-fishes the internal shell is calcareous, and lies loosely within the mantle. It consists of a broad plate, which terminates behind in an imperfectly chambered apex. It is popularly called the "cuttle-bone." The squids or **Calamaries** (Lat. *calamus*, a pen) have a horny internal shell, which consists of a central shaft, with two lateral wings. This is termed the "pen." In the **spirulæ** the internal shell is whorled, disc-shaped, and chambered, resembling the external shell of the nautilus. The chambers are connected by a tube called the **siphuncle** (Lat. *siphunculus*, a little tube). The whorls do not touch one another.

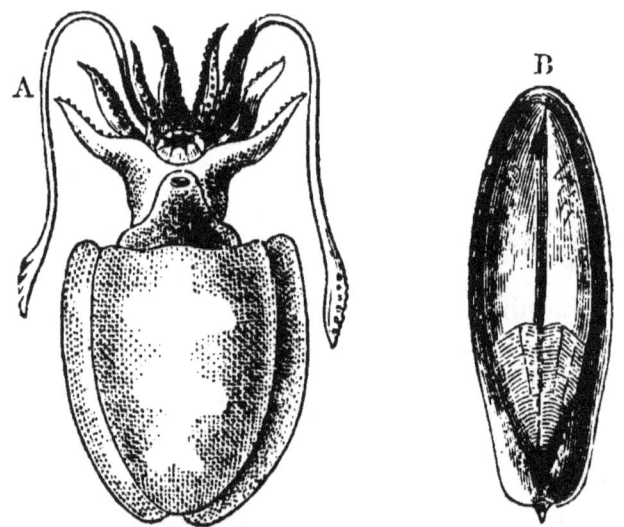

FIG. 39.—*A*, CUTTLE-FISH (VENTRAL SIDE); *B*, "CUTTLE-BONE."

The extinct **Belemnites** (Gr. *belemnon*, a dart) had an internal shell, consisting of a **phragmacone** (Gr. *phragma*, a division; *konos*, a cone), a conical chambered shell, which was lodged in a fibrous sheath, called "the guard." This "guard" is the part usually found in the rocks, preserved as a fossil.

The only living cephalopods, provided with external shells, are the pearly nautilus *(Nautilus Pompilius)*, and

MOLLUSCA—CEPHALOPODA.

the paper nautilus *(Argonauta)*. The shell of the paper nautilus is spiral, but without chambers. It is secreted by the two webbed arms of the female, but is not attached to the body of the animal. The male is much smaller than the female, and is destitute of a shell. The shell of the pearly nautilus is secreted by the mantle. It is a spiral shell, and is divided into chambers by partitions called **septa** (Lat. *septum*, a partition), which are shaped somewhat like saucers, and are pierced in the centre by a

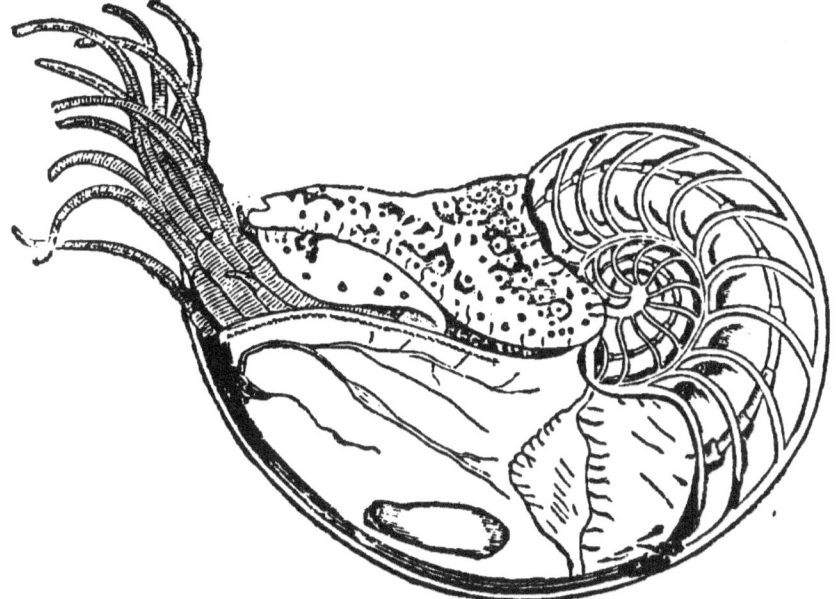

Fig. 40.—Section of a Pearly Nautilus, showing the chambers traversed by the siphuncle. The animal occupies the last of these chambers.

membranous siphuncle. As the animal increases in size, it removes to a wider part of the shell, and walls off the part previously occupied by a partition, always occupying the last formed and largest chamber. Each chamber has therefore been occupied by it in succession, and it still keeps up a connection with the vacated chambers by the siphuncle.

There are many extinct families of cephalopods, allied to the nautilus, whose remains have been deposited in the rocks from the earliest period. The best known of these is the ammonite, so abundant in the secondary rocks.

The form of the body is symmetrical, the two sides being alike. The head is distinct, and is surrounded by a circle of arms. These are usually provided with rows of suckers which act like cupping glasses. The poulpe has eight arms which are all alike long. The cuttle-fish has ten arms. Two of them, called **tentacles**, are much longer than the others, and are only provided with suckers at the extremities. Two of the arms of the paper nautilus are provided with webs which secrete the shell. The pearly nautilus has numerous arms, which are destitute of suckers. The arms are considered to be modified expansions of the part which corresponds to the foot in other molluscs.

130. Digestion.—The mouth leads into a cavity which contains two horny jaws and a **tongue**. The jaws work vertically, and bear some resemblance to the beak of a parrot. The **tongue is** partly covered with recurved siliceous spines. It has been called by Professor Huxley an **odontophore,** or "tooth bearer." "It consists essentially of a cartilaginous cushion, supporting as on a pulley, an elastic strap, which bears a long series of transversely disposed teeth. The ends of the strap are connected with muscles attached to the upper and lower surface of the hinder extremities of the cartilaginous cushions; and these muscles, by their alternate contractions, cause the toothed strap to work backwards and forwards, over the end of the pulley formed by its anterior end. The strap consequently acts, after the fashion of a chain-saw, upon any substance to which it is applied, and the resulting wear and tear of its anterior teeth are made good by the incessant development of new teeth in the secreting sac in which the hinder end of the strap is lodged. Besides the chain-saw-like motion of the strap, the odontophore may be capable of a licking or scraping action as a whole." (*Huxley.*) The cavity of the mouth is succeeded by a gullet which leads to a stomach and intestine. The

MOLLUSCA—CEPHALOPODA.

gullet is provided with salivary glands. The intestine opens on the ventral surface near the base of the **funnel**. This **funnel** is a muscular tube, situated under the head. It communicates at one extremity with the cavity of the mantle, and at the other with the external medium. The naked cephalopods are provided with a gland called the **ink bag**, which opens into the gill chamber near the anus. It discharges its contents through the funnel, and darkens the water, thus covering their retreat when pursued by an enemy.

131. Circulation.—The blood is colourless. The heart contains two cavities, an auricle and a ventricle. The

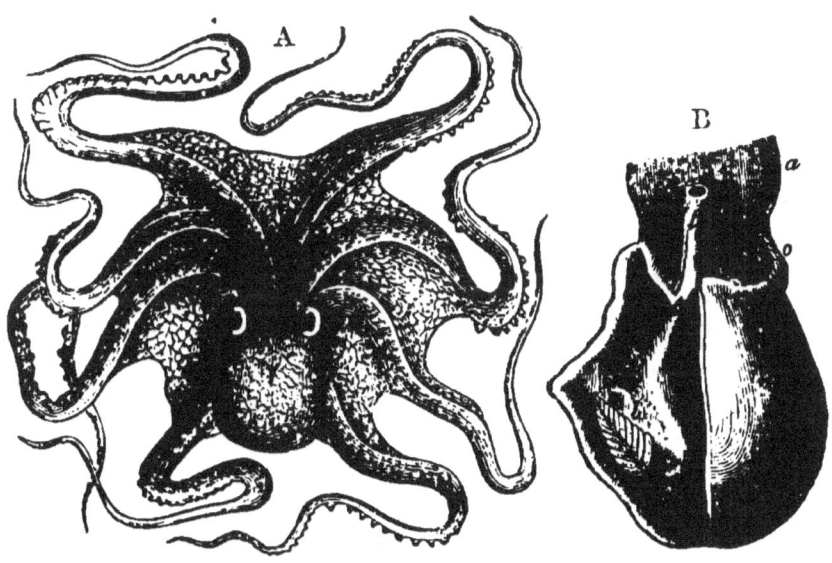

FIG. 41.—*A*, OCTOPUS OR POULPE; *B*, THE MANTLE OF THE POULPE REMOVED FROM THE RIGHT SIDE.

b, the right gill; *o*, the free margin of the branchial sac; *t*, the funnel.

aerated blood from the gills passes into the auricle, thence to the ventricle which drives it through the body. In the cuttle-fishes, at the base of each gill, there is a special cavity which by its contraction drives the venous blood that has come from the various parts of the body through these organs. These cavities are called **branchial hearts.** They are wanting in the nautili. In the cuttle-fishes, &c., the veins and arteries are connected by capillaries. It is worthy of note, that while the blood in the heart of a fish is venous, the heart of a mollusc contains arterial blood.

132. Respiration.—All cephalopods breathe by gills or branchiæ. These are plume-like organs, placed on the sides of the animal in a cavity formed by the mantle which opens in front on the under surface of the body. When the mantle expands, water is forced into this cavity; when it contracts, it is ejected through the funnel, which is provided with a valve, allowing water to pass towards the head, but not otherwise. The reaction produced by the escape of this water causes the animal to move in the opposite direction, that is, tail foremost. Thus, in the cephalopods, as in many other animals, the functions of locomotion and respiration are intimately associated.

The cuttle-fish tribe and the paper nautilus are each provided with two gills; the pearly nautilus with four.

133. The nervous system consists, as in the other Mollusca, of three ganglia united by nervous cords—the *cerebral, pedal,* and *parieto-splanchnic* ganglia. The cerebral ganglia are large, and are surrounded by a cartilaginous ring which may be considered a rudimentary skull. This is the nearest approach to an internal skeleton found in any invertebrate animal. There are two large eyes situated on the sides of the head.

134. Development.—The Cephalopoda are unisexual animals. Their ova are of large size comparatively. In the cuttle-fishes one of the arms of the male is adapted for conveying to the gill chamber of the female the sperm-cells which have previously been formed into a mass by a glutinous substance. In the common octopus or poulpe, the third arm on the right side has at first the appearance of a small bladder. When this bladder bursts, the arm is liberated. It is now considerably longer than the others, and is provided with an oval plate at the extremity, which is believed to be used, like the modified palpi of the male spiders, in conveying sperm to the female.

In the paper nautilus, this modified arm is detached from the body of the male, and lodged in the mantle chamber of the female, a fresh arm being from time to time developed. When first discovered in this position, it was believed to be an internal parasite, and received the name of *hectocotylus* (Gr. *hekaton*, a hundred; *kotulos*, a cup), because of its numerous sucking cups.

Class 2.—Pteropoda.

135. The **Pteropoda** (Gr. *pteron*, a wing; *poda*, feet) are so called because their organs of locomotion consist of two fin-like expansions of the foot *(epipodia)* placed on each side of the head. They are very small animals. They swim near the surface, sometimes in vast numbers, in the deeper parts of the ocean. The *Clio Borealis*, a naked species, not more than an inch in length, forms the principal food of the Baleen whale.

Professor Huxley places the *Dentalium*, which is common enough on our own coasts, among the *Pteropoda*.

136. **Skeleton.**—The body is covered with a mantle, which generally secretes a univalve shell of a glassy substance. This shell sometimes consists of two pieces united at the

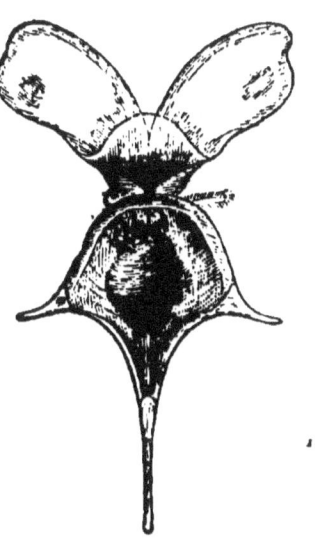

Fig. 42.—Hyalea.

edges, and is often of a spiral form, not unlike a small nautilus. Some species are destitute of a shell.

137. Digestion.—There usually is no distinct line of separation between the head and the body. The mouth is surrounded by tentacles, and has an odontophore. There is a stomach, liver, and intestine, with anal opening, on the ventral surface of the body. They are said to subsist on small Crustacea, &c.

138. Circulation.—There is a heart with two cavities, an auricle, and a ventricle.

139. Respiration is effected by a portion of the surface which is covered with cilia. This is sometimes placed in a chamber within the mantle, and is sometimes external.

140. The **Nervous System** consists of three pairs of ganglia. The principal mass is *below* the œsophagus.

141. Development.—All the *Pteropoda* are hermaphrodite, male and female organs being united in each individual. The young undergo metamorphosis.

Class 3.—Gasteropoda.

142. The **Gasteropoda** (Gr. *gaster*, the belly; *poda*, feet) is a very large class. It includes land-snails, sea-snails, fresh-water snails, slugs, limpets, whelks, &c. It is so called, because the organ of locomotion generally consists of a flat muscular disc occupying the lower part of the body, which is termed the "foot." By the alternate contraction and expansion of this organ they glide slowly along. In the *Heteropoda* (Gr. *heteros*, diverse; *poda*, feet), the foot is furnished with fin-like expansions, which adapt the animal for swimming. The hinder part of the foot, in many of the *Gasteropoda*, is provided with a horny or calcareous plate, called the **operculum** (Lat. *operculum*, a lid). When the animal retreats within the shell, which it is enabled to do by the retractor muscle, this operculum closes up the aperture.

MOLLUSCA—GASTEROPODA. 103

143. Skeleton.—The *Gasteropoda* are provided with a mantle which encloses the body like a sac. This mantle, in most cases, secretes a shell, which is generally **univalve**, or composed of a single piece. A few have **multivalve** shells, which are composed of eight pieces, placed one behind another. None possess a **bivalve** shell. Some.

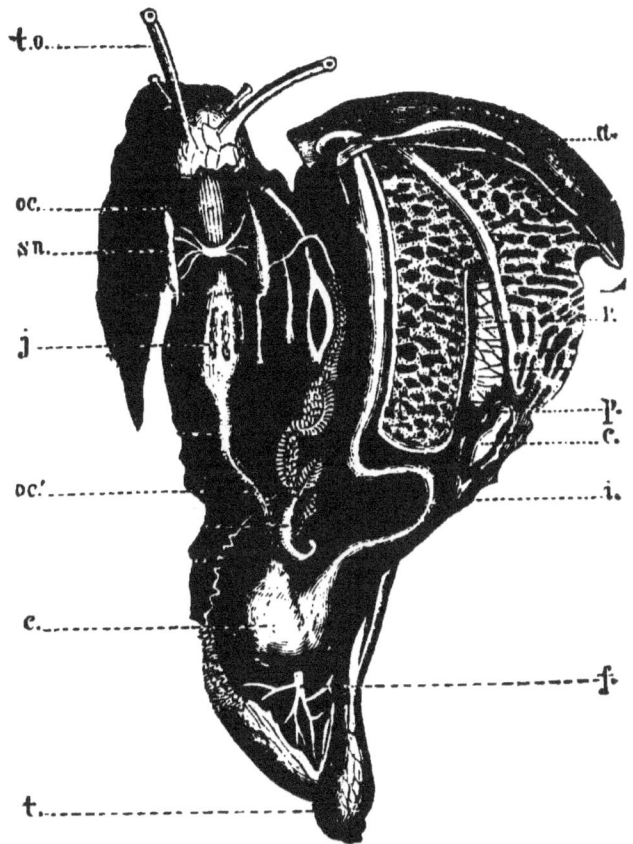

Fig. 43.—Anatomy of the Snail.

to, ocular tentacles; *oe*, œsophagus; *sn*, cerebral ganglion; *j*, gizzard; *e*, stomach; *t*, liver; *f*, biliary ducts; *i*, intestine; *a*, anus; *r*, kidney; *p*, breathing organ; *c*, heart.

as the common slugs, are destitute of shells. The univalve shell is either a simple cone, as in the *limpet;* or "spiral," which is its most usual form. The various coils of the spiral are called **whorls**. The last whorl is the largest, and is termed the **body whorl**. The part of the shell above the body whorl is the **spire**. The **sutures** are the lines of junction of the whorls. Sometimes the whorls do not form a spire, but are in the same plane, forming a disc, as in the fresh water shell *Planorbis*. In the vegetable-feeding gasteropods, the aperture or mouth of the shell is unbroken or "entire." These are therefore called **holostomata** (Gr. *holos*, whole; *stoma*, mouth). In the carnivorous species, the mouth of the shell is notched or drawn out into a canal for the protection of a breathing siphon. Sometimes there are two of these canals or notches. These breathing tubes have no immediate connection with the carnivorous habits of the animals.

144. **Digestion.**—The head is very distinct, and is usually furnished with two long tentacles. There are also often two stalks, on the summit of which the eyes are placed. The mouth is sometimes furnished with two horny jaws. There is always an **odontophore**, armed with siliceous teeth, similar to that possessed by the Pteropoda and Cephalopoda. The mouth leads to a gullet, which is succeeded by a stomach and convoluted intestine. In vegetable feeders, the stomach is often provided with calcareous plates, which assist in grinding the food. The anus is placed on the right side, in the front part of the body. There are salivary glands and a liver.

145. **Circulation.**—There is usually a distinct heart, with two cavities—an auricle and a ventricle.

146. **Respiration** is variously effected in the different groups. The land snails and slugs, and some fresh water gasteropods (*Limnæa*, &c.), breath air directly by means of a pulmonary chamber hollowed out on the right side of the body near the head. This chamber is provided with an opening to admit the air, and its walls are lined with a membrane containing numerous blood-vessels. This group of air-breathers is placed by Professor Huxley in a separate class—the *Pulmo-Gasteropoda*.

Those that breathe air through the medium of water

(Branchio-Gasteropoda) perform the respiratory function in three different ways—

1. Some of the *Heteropoda* have no special breathing organs. The walls of the mantle-cavity are supplied with blood-vessels, and the blood in passing through these is purified by the air contained in the water.

2. The sea-slugs, &c. have **external** breathing organs, consisting of tuft-like expansions of the integument placed on the back and sides of the animal. These are called *Nudibranchiata* (Lat. *nudus*, naked; Gr. *branchia*, gills), and may be found on sea-weeds or under stones on the sea-shore.

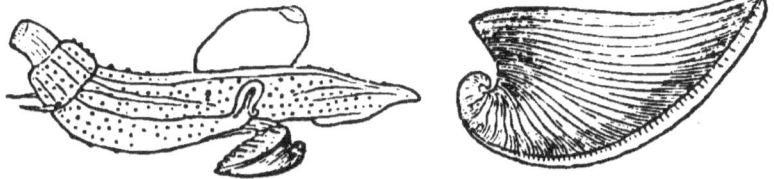

FIG. 44.—CARINARIA AND ITS SHELL *(Heteropoda.)*

3. In the majority of cases, the gills are plume-like organs, contained in a chamber, formed by a fold of the mantle. In many members of this group, the water is conveyed to the gill chamber through **a siphon**, formed by a prolongation of the mantle fold. This water passes off through an aperture which is often furnished with another siphon.

147. The **Nervous System** consists of three pairs of ganglia—the cerebral, pedal, and parieto-splanchnic. These are united by nervous threads. There are two eyes, often placed at the summit of a pair of stalks. In the land snails, these stalks are tubular, and contain muscular threads, which, by their contractions, pull the eyes into the interior of the tubes, when the animal is apprehensive of danger to these important organs. There are two auditory vesicles, filled with fluid, and containing **otoliths** (Gr. *ous*, an ear; *lithos*, a stone), which are in a constant state of vibration. These organs are placed at the bases of the tentacles.

148. Development.—The sexes are usually distinct, but are united in the sea-slugs. The young, when hatched, have a shell in all cases. The naked molluscs soon cast this off. The young of the water-breathing species are furnished with a pair of ciliated fin-like expansions, resembling the **wings** of the *Pteropoda*, by which they are enabled to swim with great facility. This is one of many instances, in which the larval form of one group resembles the permanent form of another.

Class 4.—Lamellibranchiata.

149. The **Lamellibranchiata** (Lat. *lamella*, a plate; Gr. *branchia*, gill) are so called, because they are provided with two gills or branchiæ on each side of the body, in the form of **membranous plates** or leaves. They are also called **Conchifera** (Lat. *concha*, a shell ; *fero*, I bear), and sometimes **Acephala** (Gr. *a*, without ; *cephale*, the head). They are always provided with bivalve shells. These may be distinguished from the bivalves of the *Brachiopoda*, by being usually **equivalve** and **inequilateral**; whereas, the brachiopod bivalves are always **equilateral**, and never **equivalve**. They have a mouth, but no distinct head. The **lingual ribbon,** or **odontophore,** found in the higher molluscs, is absent.

The *Lamellibranchiata* is a very numerous class, including all the bivalve **shell-fish** ordinarily met with. They are all aquatic. Most of them inhabit the sea, but a few live in fresh water. Mussels, cockles, oysters, and scallops are familiar examples.

150. Skeleton.—The mantle, instead of surrounding the animal like a sack, as in the Gasteropoda, is divided into two halves, or lobes, which are placed on the right and left sides of the body. Sometimes the mantle lobes are unconnected ; at other times they are united, except at two points, where there are openings for the passage of the foot and the breathing siphons. Each lobe secretes a separate shell, formed of carbonate of lime. These shells are really cones with their apices turned to one side—that in which the mouth of the animal is situated. The apex, where the growth of the shell commences, is termed the **umbo**, or **beak**. The beaks are situated at the **dorsal** part of the shell. The opposite part, where the shell opens, is the **base**. The side of the shell, towards which the beaks point, is called the **anterior** side, because the mouth of the animal is placed near it. The opposite is the **posterior** side. The **length** of the shell is measured from the anterior to the posterior side; the **breadth** from the beaks to the base. The anterior side is usually shorter than the posterior; the shell is therefore said to be **inequilateral**. The valves are generally equal in size; that is, they are **equivalve**. The shells of the scallop and oyster, however, are **inequivalve**. As the mouth is situated near the anterior margin, the valves are consequently placed on the sides of the animal, and are therefore termed **right** and **left**. In the Brachiopoda, as already noticed, the valves are placed on the **dorsal** and **ventral** surfaces of the body, and are always **equilateral** and **inequivalve**.

Near the beaks is a line called the **hinge line**, where the valves are united together by a series of projections called **teeth**, which fit into corresponding cavities in the opposite valves. The valves are bound together by a structure termed the **ligament**, of which there are usually two parts—one external, the other internal. This ligament is placed immediately behind the beaks. The external ligament is a fibrous horny substance, which is always stretched when the valves are shut. The internal ligament, or **cartilage**, consists of elastic fibres, which are placed perpendicularly between the two valves. When the valves are shut, these fibres are compressed and shortened. When the pressure is removed, the ligaments open the valves by their elasticity.

The valves are **shut** by the **adductor** (Lat. *ad*. to; *duco*, I lead) **muscles**. There are sometimes two of these muscles, and sometimes only one. One is placed in the anterior part of the shell, and one in the posterior. The anterior muscle is sometimes absent. These muscles leave **impressions** in the interior of the valves, so that it can be easily ascertained, by examining a shell, whether the living animal had one or two. The valves in ordinary circumstances are kept open; but, when the animal is alarmed, it exerts its adductor muscles, and closes them. A dead bivalve always gapes, because, when the adductor muscles cease to act, the elasticity of the ligament opens the valves.

FIG. 45.—VENUS, showing Foot and Breathing Siphons.

Most lamellibranchs possess a **foot**, which is protruded through an opening in the mantle, on the lower surface of the body. The foot is a muscular organ, sometimes finger-shaped, and sometimes bent like a sickle. This foot enables the animals, in some instances, to move along by making a number of short leaps. In other cases it is used in making burrows in the sand or mud. In the mussels, &c., the foot is furnished with a gland which secretes a silky substance. Of this **liquid silk**, the creature forms a tuft of thread, by which it attaches itself to a stone or some other object. The foot is sometimes very small. In the oyster tribe it is altogether absent.

The animal is provided with muscles, by which it is enabled to protrude or retract the foot at will. These **retractor muscles of the foot** leave certain markings in the interior of the shell, which are called the **pedal impressions.**

There are other "impressions" in the interior of the shell, of which the principal are the **pallial line,** and the **pallial sinus.**

The **pallial** (Lat. *pallium,* a cloak) **line** marks the impression made by the margin of the mantle.

The **pallial sinus** (Lat. *sinus,* a bay), is an indentation in the pallial line, found in those species that have retractile breathing siphons. The depth of the indentation corresponds with the length of the siphons.

151. Digestion.—The *Lamellibranchiata* are destitute of a distinct head. The mouth is placed in the front part of the body, and is furnished with one or two pairs of membranous organs called **palpi.** There are no teeth. The mouth is followed by a gullet which conducts to a stomach and convoluted intestine. The intestine passes through the heart, and terminates in an anus, placed near the aperture of the breathing siphon. There is a large liver, but no salivary glands. Their food consists principally of small animalcules.

152. Circulation.—There is a well developed heart, consisting often of three chambers—two auricles and one ventricle. Sometimes there is only one auricle and one ventricle, as in the oyster. In a few instances there are two hearts, placed in different parts of the body, each composed of an auricle and a ventricle. There are arteries which convey the blood from the ventricle through the body, and veins which carry it to the gills, where it is aerated. From these it is brought back to the auricle or auricles.

153. Respiration.—The organs of respiration consist of a pair of plate-like gills on each side of the body.

110 ZOOLOGY.

These gills are attached along their dorsal margins. Sometimes there is only one gill on each side. When viewed by a microscope, these gills are seen to consist of a number of minute "tubular rods supporting a network of blood-vessels." They are covered with cilia, which, by their perpetual movements, keep up a constant current of water. This current not only supplies fresh water for the purification of the blood,

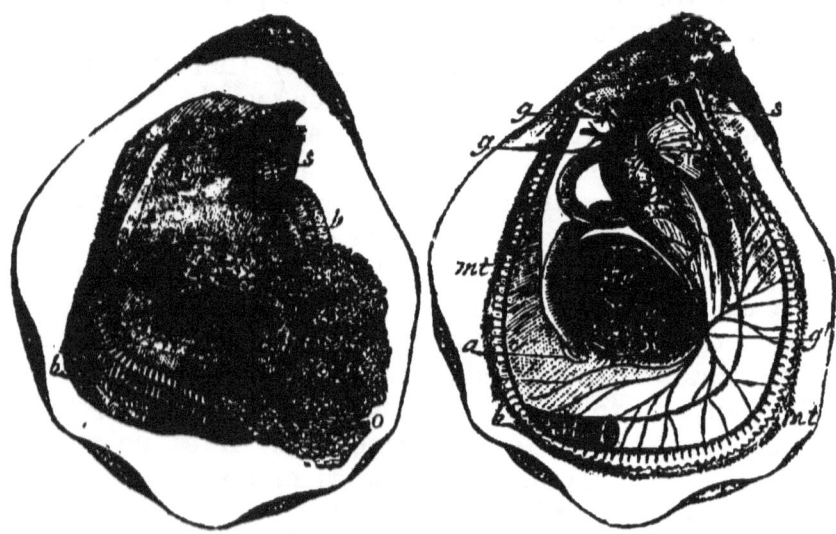

FIG. 46.—ANATOMY OF THE OYSTER.

s, mouth; *e*, stomach; *i i*, intestinal canal; *a*, anus; *b g*, nervous ganglia; *m t*, mantle; *b* (figure to the left), branchiæ; *m*, adductor muscle; *o*, ovary.

but also brings minute particles of food for the sustenance of the animal.

Those lamellibranchs that live buried in sand or mud are provided with two respiratory siphons or tubes. One of these admits a current of water which passes over the gill plates, supplies the mouth with such food as the water may contain, and is then discharged through the other siphon, carrying with it the excretions from the intestine. These siphons are sometimes longer than the shell, and in other cases are very short. When the animal buries itself, it always leaves the extremities of the siphons exposed at the surface. The siphons are sometimes separated from one another; in other instances they are united along one side, forming a compound tube. The animal has generally the power of withdrawing the siphons within the shell. The muscle by which this is effected is termed the **retractor muscle of the siphons.** The indentation produced in the pallial line by the action of this muscle has already been described.

154. The **Nervous System** consists of the three normal pairs of ganglia—the **cerebral** pair, placed near the mouth; the **pedal** pair in the foot; the **parieto-splanchnic** pair, which supplies filaments to the mantle, gills, and other internal organs. These ganglia are connected by nervous cords.

Eyes are present in the form of minute coloured spots which are "set at intervals like a row of sentinels," round the borders of the mantle.

Auditory sacs, containing otoliths, are placed near the pedal ganglia.

155. **Development.**—The sexes are generally separate, but are united, in some cases, in the same individual. The ova are retained within the body of the parent until they are hatched. When set free, they are able to swim, being furnished with cilia. Thus, the fry or **spat** of the oyster can move about, while the adult animal is sessile.

SUB-KINGDOM V.—**Molluscoida.**

CLASS 1.—Brachiopoda.

156. The **Brachiopoda** (Gr. *brachia,* arms; *poda,* feet) are so called, because they are provided with two long arms which are coiled up within the shell. They are all marine animals; and, as most of them inhabit the deep seas, they are less known than the shells of the true molluscs are. They are often called **lamp-shells,** as many of them have a striking resemblance to an ancient

lamp. They have a very wide range, being found in all latitudes, and at various depths of sea bottom. They are much more commonly met with as fossils than as recent shells, nearly two thousand extinct species having been described, while less than one hundred living species are known. They are found in all the geological formations from the Cambrian upwards. They are especially abundant in the Silurian rocks.

FIG. 47.—ANATOMY OF TEREBRATULA (the animal withdrawn within its shell).

a a, mantle ; *c c*, peduncle ; *d d' d''*, muscles which open and close the valves ; *e*, stomach ; *i i*, intestine ; *k*, pseudo-heart ; *n n*, ciliated arms ; *o*, oviduct.

157. Skeleton.—The Brachiopoda, like the Lamellibranchiata, possess bivalve shells. The valves of the brachiopod shells are always more or less unequal in size. The shells are therefore said to be **inequivalve**. They are **equilateral**—that is, the two sides of the valves are equal and symmetrical. On the other hand, the valves of the Lamellibranchiata are generally equal in size, but are always inequilateral and unsymmetrical. The valves are **dorsal** and **ventral**, instead of being **right** and **left**, as in the Lamellibranchiata. The ventral valve is always the larger.

The valves are often united together, along a hinge

line, by two teeth attached to the ventral valve, which are inserted in sockets in the dorsal valve. There is no ligament, the valves being both opened and closed by the action of special sets of muscles. The shells are generally calcareous, but in the Lingula they are of a horny texture.

The ventral valve is usually provided with a long beak which is often perforated to allow the passage of a muscular stalk, called the **peduncle** (Lat. *pedunculus*, a stalk), by which the animals are attached to rocks, corals, &c. The **Terebratula** (Lat. *a little bore hole*) derives its name from this aperture. This peduncle sometimes passes out between the valves, and in other cases the animal is attached by the beak of the ventral valve, the peduncle being absent. It is believed that some fossil

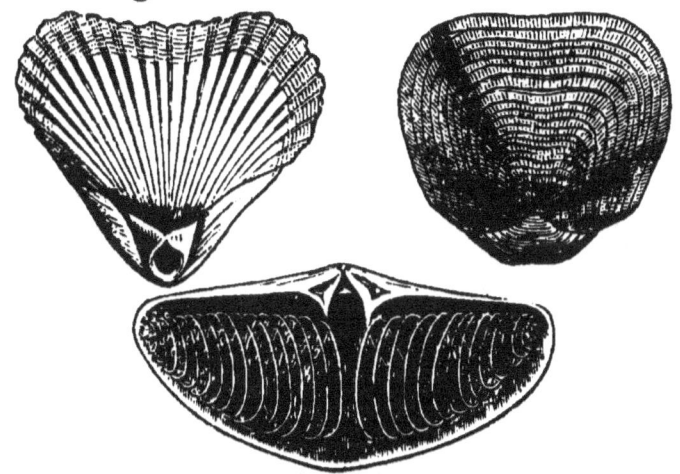

FIG. 48.—FOSSIL BRACHIOPODA.
Rhynchonella—Productus—Spirifer.

groups were able to move about freely. The valves are lined by the mantle lobes. The interior of the shell usually contains a calcareous skeleton, attached to the inner surface of the dorsal valve, upon which the arms are supported. In the extinct *spirifers*, this skeleton was coiled up like a pair of watch springs. In the *Terebratula* it has the form of a loop.

158. Digestion.—The digestive organs are confined to a very small space near the beak of the shell. This space is separated from the rest of the interior cavity by a membrane, in the centre of which the mouth is placed. The mouth is furnished with two arms, which correspond to the labial tentacles of the Lamellibranchiata. It was formerly thought that the animal had the power of protruding these arms, but it is now believed that they remain permanently coiled up within the shell. They are often three or four times the length of the shell, and occupy a large portion of the interior cavity. They are furnished on one side with a fringe of filaments, called **cirri** (Lat. *cirrus*, a curl). These cirri are provided with microscopic cilia, which produce a current in the water, and thus bring a supply of food.

The mouth leads to a gullet and stomach, which is surrounded by a large liver. The intestine in some cases terminates in a blind sac; in others it is provided with an anal opening. It is always completely shut off from the cavity of the body.

159. Circulation.—The heart is in the form of a simple sac, without any valve or partition, and is placed on the dorsal side of the stomach. The course of the circulation has not been fully ascertained.

These animals possess what Professor Huxley has termed an **atrial** (Lat. *atrium*, a hall) system, from its resemblance to the **atrial chamber** in the Ascidians. It consists of a series of canals which communicate with the exterior by two or four organs, called **false hearts,** because they were formerly believed to be connected with the blood vascular circulation. The object of the system seems to be to carry off excretions, and the products of the generative organs.

160. Respiration was formerly thought to be effected solely by the mantle; hence the name *Palliobranchiata* (Lat. *pallium*, a cloak ; Gr. *branchia*, a gill) was given to this class. The long fringed arms are now believed to discharge the office of purifying the blood.

161. The **nervous system** consists of a single principal ganglion, which, in those groups that have the valves united by a hinge, is connected with a chain of ganglia passing round the gullet.

162. **Development.**—The sexes are sometimes distinct; in other cases they appear to be united in the same individual. The young are believed to be able to swim about by the action of their fringed arms.

163. **Divisions.**—They are divided into two groups—the *Articulata* and *Inarticulata*. In the Articulata the valves are united along the hinge line by teeth and sockets. The intestine ends in a blind sac. The mantle lobes are united dorsally. The *Terebratula* belongs to this division. In the *Inarticulata* the valves are not united by teeth. The intestine is provided with an anal opening. The mantle lobes are not connected. The *Lingula* is an example of this group.

Class 2.—Tunicata, or Ascidioida.

164. The **Tunicata** (Lat. *tunica*, a cloak) are so called because they are covered with a leathery integument. They are also called **Ascidioida** (Gr. *askos*, a wine bag), because many of them are shaped like a double-necked bottle. When touched, they discharge a jet of water; on this account they have received the popular name of **sea-squirts**. They are all marine animals, and may generally be found on our coasts attached to sea-weeds, or shells, especially after a storm. They are easily known by their two prominent apertures.

Some of the Ascidians are single, or **simple**. These are fixed to some solid object, such as a submerged stone. The **social** Ascidians form a colony by budding. They are connected by a tube, through which the blood common to the colony flows. In the **compound** Ascidians the tests are joined together into a mass, but there is no common circulation. Other groups float about on the surface of the sea. Some of these, as the **Pyrosoma** (Gr. *pur*, fire; *soma*, a mouth), are phosphorescent, and light up the sea in tropical regions with a pale greenish light.

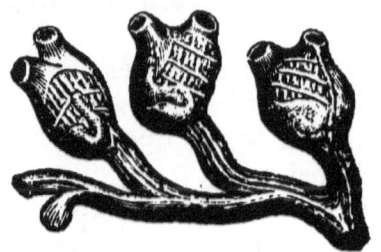

Fig. 40.—Social Ascidians.

165. Skeleton, or Covering.—In the simple Ascidians the covering consists of two coats, an outer and an inner. The outer coat is rough and leathery in texture. It is called the **test or external tunic**. It is largely composed of **cellulose**, a chemical substance closely allied to the woody fibre of plants. This substance is rarely found in animal tissues. The inner integument, or mantle, is muscular. It is by means of its contractability that the animal is able to squirt out water through one of its apertures.

166. Digestion.—The two orifices are usually guarded by short tentacles, which are not furnished with cilia. The upper or oral opening leads into a chamber which occupies a large portion of the interior of the animal. This is usually called the **respiratory** or **branchial** sac. Its walls are perforated like a sieve, the apertures being lined with cilia. At the lower extremity of this sac the entrance to the gullet is situated. The gullet leads to a stomach and intestine, which is bent round, so as to reach a cavity connected with the other aperture, called the **cloaca** or **atrial** (Lat. *atrium*, a hall) **chamber**. A current of water constantly passes in at the oral aperture into the respiratory chamber, and through its perforated walls to the atrial chamber. This current is produced by the movements of the cilia which line the apertures of the respiratory sac, the cilia working so as to produce a current in that direction. From the atrial chamber the water is discharged through the **atrial aperture** to the surrounding medium, having in its passage not only brought food to the stomach, but also carried off the excretions supplied by the intestine.

167. Circulation.—The heart is in the form of a simple muscular tube, open at both ends, and without valves. By the peristaltic motions of this tube circulation is effected. For a definite number of times the blood is propelled in one direction; then, the action of the tube being reversed, it flows as many times in the opposite direction. Thus the two ends of this tubular heart perform alternately the functions of an artery and a vein.

168. Respiration.—The walls of the respiratory chamber are richly supplied with blood-vessels. The blood in these vessels is aerated by the air contained in the water which constantly flows through it. This current of water not only brings food to the animal, but also brings oxygen for the purification of its blood. Thus, in the ascidians, as well as in the Mollusca proper, the functions of respiration and alimentation are intimately associated.

169. The nervous system consists of a single ganglion placed between the oral and atrial apertures. There are generally eye-like spots between the tentacles that surround the mouth. There is also an auditory vesicle containing an otolith.

170. Development.—The ascidians are all hermaphrodite, the sexes being united in the same individual. The young in some groups resemble tadpoles, being free-swimming, and provided with long tails.

Class 3.—Polyzoa.

171. The **Polyzoa** (Gr. *polus*, many; *zoon*, an animal) is the lowest class of the Molluscoida. It includes the "sea-mosses" and the *Flustra* or **sea-mats**. These animals were formerly placed along with the Hydrozoa, which they very much resemble in outward appearance; their internal structure, however, is widely different. From their resemblance to plants, they are often called **Zoophytes** (Gr. *zoon*, an animal; *phuton*, a plant). Most of them inhabit the sea, but some species live in fresh water. They are compound animals, forming colonies which have been produced by **gemmation** (Lat. *gemma*, a bud), or budding from a single individual. These colonies are provided with a chitinous or calcareous integument.

Fig. 50.—Flustra.

The Flustra may be readily found covering the large fronds of the Laminarian sea-weed with a lace-like crust. Other species are spread over shells or submerged stones. Others look like diminutive shrubs, and are merely attached by one extremity to some solid object. They are so very minute that 1,800 cells, each containing a separate animal, have been found in one square inch of surface.

To the outside of some of the cells certain organs are often attached which bear a considerable resemblance to the beak of a bird. They have been called **avicularia** (Lat. *avicula*, a little bird), or "bird's head processes." The beak consists of two **mandibles**, the lower of which only is movable. This organ is quite independent of the animal to which it is attached, for it continues its snapping movements long after the death of the zooid. The use of these curious appendages is not known. They are supposed to bear some analogy to the **pedicellariæ** of the sea-urchins and star-fishes.

A colony of these animals is called a **polyzoarium**, each individual animal a **polypide**. The term **zooid** is used both in Hydrozoa and Polyzoa, as a common designation for the individual organisms that make up a colony which has been produced by gemmation or budding.

172. Skeleton.—Each polypide is enclosed in a double-walled cell. The outer layer is either chitinous or calcareous, and is called the **ectocyst** (Gr. *ektos*, without; *kustis*, a bladder). The inner lining of the ectocyst is membranous, and is called the **endocyst** (Gr. *endon*, within; *kustis*, a bladder). There are two openings in this membrane, one for the mouth and the other for the anus. These openings are generally placed near one another.

173. Digestion.—The mouth is surrounded by a number of hollow tentacles which are provided with cilia. In the marine species, these tentacles are arranged in the form of a circle; in those that live in fresh water, they assume the shape of a crescent or horse-shoe. The cilia, by their perpetual movements, produce little whirlpools in the water which supply the animals with those nutritious particles on which they subsist.

When the animal is alarmed, it is able by a special set of muscles, connecting the lower wall of the cell with the gullet, to retract the front part of the alimentary tube with its attached crown of tentacles, as well as the anterior part of the membranous sac, within the posterior part of this sac, by a process of inversion, somewhat similar to that by which a snail withdraws its eye-stalks.

The mouth leads to a gullet and stomach, a muscular gizzard sometimes intervening. The intestine is bent upon itself, so that the anal opening is placed near the mouth. The whole perivisceral cavity is filled with fluid.

174. Circulation.—No heart has yet been discovered in any *Polyzoon*. The products of digestion pass through the walls of the intestine, and become mixed with the surrounding fluid.

FIG. 51.—PLUMATELLA.

a, branchial plume; *b*, œsophagus; *c*, stomach; *d*, intestine; *e*, anus; *f*, egg suspended from the ovary.

175. Respiration appears to be effected by the ciliated tentacles, probably assisted by the fluid which fills the perivisceral cavity. This fluid is slowly replaced by the action of the cilia which line the interior walls of the endocyst.

176. Nervous System.—Between the mouth and the anal aperture there is placed a single ganglion, which sends off nerves in different directions. There is also said to be a "colonial nervous system," forming a means of communication between the various polypides in a colony. It is believed to be by means of this nervous system that the *avicularia* are able to continue their motions after the death of the zooids.

177. Development.—In the Polyzoa the sexes are united in the same individual. The reproductive organs are attached to the inner surface of the endocyst. The ova pass into the perivisceral cavity, where they are fertilized. How they make their escape from the cell has not been ascertained. The embryo is at first able to swim about by the action of its cilia. After a time it attaches itself to some fixed object, and originates a new colony, the number of zooids being increased by continuous budding.

QUESTIONS—IV.

1. What is the derivation of the term *Cephalopoda*?
2. Give examples of this class.
3. What are its distinctive characters?
4. What is the habitat of these animals?
5. What are their habits?
6. Describe the "arms," the "funnel," and the "mantle."
7. What are the distinctive features of the arms in the Poulpe, the Cuttle-fish, the Paper Nautilus, and the Pearly Nautilus?
8. How is locomotion effected?
9. What cephalopods possess an external shell?
10. Describe the external shell, and show how it differs from the univalve shell of the *Gasteropoda*.
11. What members of this class have an internal shell?
12. What cephalopods are destitute of a shell?
13. What is the nature of "cuttle-bone"?
14. Describe the "pen" of the *calamary*.
15. Describe the internal shell of the *Spirulæ*.
16. What are *Belemnites*?
17. What are *Ammonites*?
18. What is the nature of the digestive apparatus in this class?
19. Describe the "jaws" and the *odontophore*.
20. Describe the "ink-bag," and state whether it is possessed by all the members of this class.
21. Describe the heart and circulatory organs.
22. What are "branchial hearts"?
23. What cephalopods are destitute of these organs?
24. Does the heart contain venous or arterial blood?
25. Describe the breathing organs.
26. How is respiration effected?
27. Show that respiration and locomotion are associated in this class.
28. On what principle are the Cephalopoda divided into orders? Name the orders.
29. Give an account of the nervous system and the organs of sight.
30. What is a *Hectocotylus*?
31. Why are the *Pteropoda* so called?
32. Give examples of this class. What is their usual habitat?
33. Describe the shell of the *Pteropoda*.
34. What is the nature of their digestive apparatus?
35. How is respiration effected?
36. Explain the term *Gasteropoda*.

37. Give examples of this class, and state whether they live on land, in fresh water, or in the sea.
38. How is locomotion effected in this class?
39. What is meant by the "foot"?
40. What are the *Heteropoda*?
41. What is the *operculum*?
42. Describe the univalve shell possessed by most gasteropods.
43. What are multivalve shells?
44. What gasteropods are without shells?
45. How may it be ascertained, by examining a gasteropod shell, whether the inhabitant was carnivorous or vegetable-feeding?
46. Is there any connection between the notches found in certain shells and the carnivorous habits of the animals?
47. Describe the "eye-stalks" of a snail.
48. Give an account of the digestive apparatus.
49. What is the distinction between the *Pulmo-Gasteropoda* and the *Branchio-Gasteropoda*? Give examples of each.
50. How is respiration effected in the *Heteropoda* and sea-slugs?
51. What air-breathing gasteropods live in water?
52. What are *otoliths*?
53. Show that the larvæ of certain gasteropods resemble the adult *Pteropoda*.
54. Why are the *Lamellibranchiata* so called?
55. Give examples of this class, distinguishing between marine and fresh-water species.
56. By what other names is this class known?
57. Distinguish the bivalve shells of the *Lamellibranchiata* from those of the *Brachiopoda*.
58. Explain the following terms:—*Umbo, inequilateral, equilateral, equivalve, length, breadth, posterior side, anterior side, base, teeth, ligament, cartilage.*
59. How are the valves opened and shut?
60. Why does a dead oyster always gape?
61. What is meant by the "foot" of a lamellibranch?
62. Explain the terms:—*Pallial line, pallial sinus, muscular impressions.*
63. Describe the digestive apparatus.
64. What are palpi?
65. What is believed to be the nature of the food?
66. Describe the heart and circulatory organs.
67. How is respiration effected?
68. Describe the gills.
69. How are respiration and sustentation connected in this class?
70. What lamellibranchs possess breathing siphons?
71. Describe the nervous system and the organs of sight?
72. Give examples of sessile lamellibranchs.

73. By what means are the fry of the oyster removed to a distance from the parent bed?
74. Explain the term *Brachiopoda*.
75. Give examples of this class.
76. Give an account of the distribution in time of the *Brachiopoda*.
77. What is their usual habitat?
78. Describe the brachiopod shell.
79. How are the valves opened and closed?
80. What is the *peduncle*?
81. Describe the *arms*, and state what is their supposed function.
82. Give an account of their digestive apparatus.
83. What is known of the circulation?
84. What is meant by the *atrial system*?
85. What is the nature of the respiration?
86. Give an account of the nervous system.
87. Into what two groups are they divided?
88. Give an account of the *Tunicata* or *Ascidioida*.
89. Explain these terms.
90. Distinguish between *simple*, *social*, and *compound* ascidians.
91. Describe the *test*, and state what vegetable substance it contains.
92. Give an account of the *branchial sac* and the *atrial chamber*.
93. What is the nature of the circulation?
94. How is the blood purified?
95. What is the nature of the nervous system?
96. Show that the larvæ in some groups differ widely from the grown animals.
97. What are the *Polyzoa*?
98. Give examples. Why are they sometimes called *zoophytes*?
99. Give a general description of these animals.
100. How do they differ from the *Hydrozoa*?
101. What is meant by *gemmation*?
102. What are *avicularia*?
103. Explain *polypide, polyzoarium, ectocyst, endocyst*.
104. What is the nature of the digestive apparatus?
105. What is known of the circulation and respiration?
106. Give an account of the nervous system.
107. How are new colonies formed?

CHAPTER V.

Cœlenterata and Protozoa.

Sub-Kingdom VI.—Cœlenterata.

Class 1.—Actinozoa.

178. The **Actinozoa** (Gr. *aktis*, a ray; *zoon*, an animal) is the highest class of the Cœlenterata. It includes the sea-anemones, the corals, the sea-pens, and some of the jelly-fishes. These animals are exclusively marine.

The Actinozoa are defined as cœlenterate animals, in which *the mouth leads to an alimentary canal, which, instead of being continued through the body, and terminating at the exterior, communicates by a wide aperture with the general cavity of the body. This canal is separated from the body walls by a space which is divided into compartments by a series of vertical partitions called* **mesenteries** (Gr. *mesos*, the middle; *enteron*, intestine). *To the surfaces of these mesenteries the ovaries and sperm cells are attached.*

179. The body consists of two separate layers—an outer and an inner. The outer is termed the **ectoderm** (Gr. *ectos*, without; *derma*, skin), and the inner the **endoderm** (Gr. *endon*, within; *derma*, skin). The interior of the body is filled with fluid, in which a circulation is promoted by the cilia that line the inner walls of the endoderm. The only digestive organ is a stomach, in the form of a tube, which can only communicate with the exterior through the mouth. No heart or circulatory organs have been discovered in any member of this class. In one group only *(Ctenophora)* is there any trace of a nervous system. The sexes are often united in the same individual. Some are simple, others are compound. Colonies are formed by budding, or by self-division of a single individual, as

CŒLENTERATA—ACTINOZOA.

in the Polyzoa. The individual members of these colonies are called **polypes** (Gr. *polus*, many; *pous*, foot); the entire colony, the **actinosoma** (Gr. *aktis*, a ray; *soma*, a body); the common fleshy structure, the **cœnosarc** (Gr. *koinos*, common; *sarx*, flesh). The corals are fixed; the sea-anemones are capable of shifting their position very slowly; the *Ctenophora* can swim with facility by the action of their cilia. The sea-anemones and *Ctenophora* are destitute of any hard structure. Some groups secrete a horny, others a calcareous skeleton, which is called **coral**.

180. The common **sea-anemone** (*actinia*) forms a good example of this class. It is a simple animal, the *actinosoma* consisting of a single polype. It may be found on the sea shore, in almost any rocky pool. When the tide is out, it looks like a little lump of red flesh; but, when it gets covered with sea water, its true shape soon appears. It is of a cylindrical form, and is firmly, but not permanently, attached to a rock by its lower part or **base**, which acts like a sucker. It is able to move along, at the rate of a few inches a

FIG. 52.—SEA-ANEMONE, attached to a gasteropod shell.

day, somewhat after the manner of a snail, but much more slowly. Opposite the base is the **disc**, in the centre of which the mouth is placed. The circumference of the disc is surrounded by two rows of tentacles, arranged alternately in concentric circles. There are sometimes as many as two hundred of these tentacles. They are hollow tubes, communicating with the cavity of the body. The animal is able, by contracting its body walls, to force water into these tentacles, and thus protrude them to a considerable length. It has also the power of withdrawing the water, and contracting them. Sometimes the tubes are open at the tips also. These tentacles are the prehensile organs of the animal. They are furnished with those singular weapons called **thread cells.** Each of these cells consists of a membranous sac, containing a coiled up thread. When the cell is touched, a filament, barbed at the point, and serrated along the edges, is projected from it with sufficient force to penetrate the tissues of most animals with which it comes in contact. These thread cells are also found in the lining of the alimentary canal, and in the outer integument, which is of a soft leathery structure.

The mouth leads into a tubular stomach, which is not followed by an intestine, but opens below into the general cavity of the body by a wide aperture. The indigestible portions of the food are ejected out of the mouth. The alimentary tube terminates about half-way between the mouth and the base. The intervening space, between this tube and the body walls, is divided by a number of vertical partitions, or **mesenteries,** into compartments. Some of the partitions, which originate in the body wall, do not reach the whole way to the outer lining of the alimentary tube. These compartments communicate with the open bases of the tentacles. To the walls of the mesenteries the reproductive organs are attached. The sea-anemones have generally the sexes in distinct individuals. The embryo is a minute oval body, and is able to swim about freely by the action of its cilia.

181. The "reef-building corals" differ from the sea-anemone mainly in having the power of secreting within their tissues a skeleton composed of carbonate of lime. This skeleton is termed the **corallum**. In its simplest form it consists of a calcareous cup, of a cylindrical or conical

Fig. 53.—FUNGIA, a simple form of Coral, inhabited by a single Polype only.

form, with radiating lamellæ called **septa**. Some of these septa reach and others fall short of a central column which is generally present. The outer cup or **theca** (Gr. *theke*, a sheath), is secreted by the inner surface of the endoderm of the body wall; the septa, by the radiating partitions or mesenteries. The corallum is thus a cast in carbonate of lime of the polype that forms it.

The reef-building corals are generally compound animals, forming colonies by budding and self-division. New colonies are formed by minute ova, which are produced at certain seasons. Like the embryos of the sea-anemone, they are able to swim about by the action of their cilia. They are united by a common living substance, or cœnosarc. "They constitute a true republic, in which each citizen, whether consciously or not, is compelled to support the general state, by which he himself exists." (*Haughton*). The cœnosarc secretes a common calcareous structure, which unites all the cups into one strong mass. Coral, formed thus *within* the tissues of the animal, is called **sclerodermic** (Gr. *skleros*, hard; *derma*, skin). The septa of modern corals are always multiples of *five* or *six*; while, in extinct corals *(Rugosa)*, they are uniformly multiples of *four*.

182. **Coral Reefs.**—The coral polypes, small as they are, have been important rock-forming agents in all periods of the world's history. The stony masses formed by them are termed **coral reefs**. The **coral zone** extends about 30° on each side of the equator. The reefs are principally found in the Indian Ocean, on the east coast of Africa, and near the shores of Madagascar; in the Persian Gulf, the Red Sea, and the Bay of Bengal; in the Atlantic Ocean, round the West India Islands, and the coast of Florida; they are especially abundant round the islands and continents washed by the Pacific Ocean. The coral animals require a mean annual temperature of not less than 68°. On this account they are rare, and of small size, in our seas at present. No coral reefs are found either along the west coast of Africa, or the corresponding coast of South America, on account of the cold currents which flow in these places. These reefs assume three principal forms.

183. (1.) **Fringing Reefs,** which skirt the shores of continents and islands. These reefs have been constructed in water comparatively shallow, and are generally not far from land.

184. (2.) **Barrier Reefs** are distinguished from fringing reefs, by occurring at a considerable distance from the nearest land. They are also surrounded by deep water. The great barrier reef on the north-east coast of Australia, is a notable example of this class. This reef is about 1,100 miles long, and 30 broad on an average,

Its distance from the Australian coast is about 30 miles. The sea on the inner side is 300 feet deep; in some places outside it is 1,800 feet. Sometimes a barrier reef surrounds an island, which on this account is called a **lagoon island**, the intervening water being called a **lagoon**.

185. (3.) **Atolls** are rings of coral, surrounding a central lagoon, in which there is no island. They are usually pierced by several openings, and form excellent natural harbours. As the coral animal cannot subsist in fresh or brackish water, these gaps are believed to have been originated by rivers which once flowed from the island round which the polypes began to build. There are several groups of atolls both in the Indian and Pacific Oceans. The ring of coral is sometimes 90 miles in diameter, but is seldom more than a quarter of a mile broad. Although only raised a few feet above sea level, many of them have human inhabitants.

The coral animal, it is evident, could never raise a reef above the sea level, as it cannot exist out of the water. The reefs are raised by the action of the waves which detach masses of rock, and pile them up—an island in the course of ages is thus formed. The interstices of the masses are gradually filled up by calcareous sand, also washed off by the action of the waves, and the whole mass is cemented together by water containing carbonate of lime in solution.

A coral reef, as has been stated, sometimes extends downwards as much as 1,800 feet, and it was formerly believed that the coral polypes commenced to build at these depths, and gradually raised their structures until they reached the surface. Darwin has proved the incorrectness of this theory, and has shown that the coral animals cannot live at a greater depth than from 15 to 30 fathoms. He accounts for the occurrence of coral reefs at greater depths, by the gradual subsidence of the sea bottom. He believes that all coral reefs were originally constructed in shallow water, near the shore of some island or continent—that is, they were once in the state

130 ZOOLOGY.

of **fringing reefs**. The sea bottom gradually sank, but the coral animals kept pace with this subsidence, building upwards, so as to keep their structure nearly as high as the sea level. The island, surrounded by a reef, ultimately disappearing beneath the waves, the ring of coral would be converted into an **atoll**, which alone remains to show where an island had been. When a barrier reef surrounds a **lagoon island**, it is evident that if subsidence goes on the island will ultimately disappear, and the surrounding reef be converted into an **atoll**. Darwin believes, that in those areas where fringing reefs are found, the land is either rising or stationary; and where **atolls** and **barrier reefs** occur, the sea bottom is subsiding.

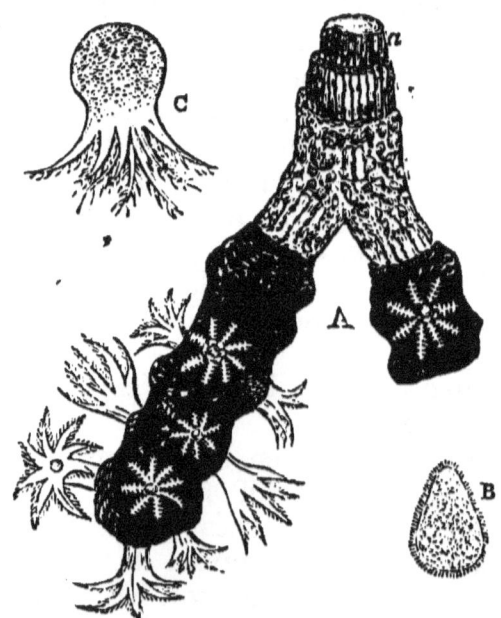

Fig. 54.—STRUCTURE AND DEVELOPMENT OF CORAL.

A, Branch of Red Coral; *a*, the calcareous axis, constituting the Red Coral of commerce; some of the polypes are expanded, and others contracted. *B*, ciliated larva of coral. *C*, larva developed, ready to fix itself and form a new colony.

186. The red coral, found in the Mediterranean and other warm seas, is an example of what is called **sclerobasic** (Gr. *scleros*, hard; *basis*, foundation) coral. This

coral is branched like a tree, and was formerly believed to belong to the vegetable kingdom. There is a rind of fleshy matter which surrounds a central stony stem, just as a tree is invested by its bark. This fleshy matter is the cœnosarc, and in it the polypes are embedded. The polypes are each furnished with eight leaf-like tentacles, which are fringed at the edges. The living rind secretes the internal skeleton, which is composed mainly of carbonate and phosphate of lime. This skeleton, although it is *within* the cœnosarc, taken as a whole, is really *outside* the polypes—that is, it is not secreted within their tissues, as in the sclerodermic corals. If a number of sea-anemones were united together into a colony, and placed round a central stem, having their bases in contact with it, and their discs outside; it is evident that, although this stem would be *within* the combined substance of the community as a whole, it would be *outside*, or at the *base* of each individual polype. In some of the sclerobasic corals the skeleton is horny, in others it consists of alternate joints of calcareous and horny matter.

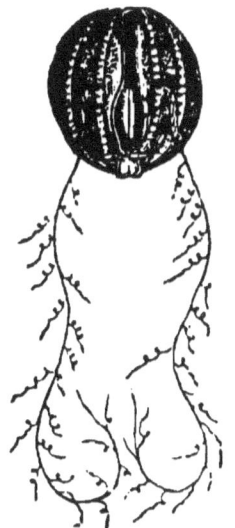

FIG. 55.—BEROE, ONE OF THE CTENOPHORA.

The tentacles are fringed, and these, as well as the mesenteries, are always some multiple of four.

Closely allied to the red coral are the **dead men's fingers,** and **virgularia** or **sea-rods,** both of which are met with in British seas.

187. The **Ctenophora** (Gr. *kteis*, a comb; *phero*, I carry) are a free-swimming, gelatinous group of actinozoons, which are popularly placed among the **jelly-fishes.** They are usually found on the surface of the ocean, far from land, swimming rapidly by the action of the cilia with which they are provided. They are very abundant in the northern seas, and form the food of several

whales. Some of the commonest forms are melon-shaped, but not bigger than a boy's marble. The cilia are arranged in eight bands, or ctenophores, extending from the upper to the lower extremity of the body. They are also furnished with a pair of long-fringed tentacles, which can be completely retracted within sacs. These animals are destitute of any skeleton or hard parts. They contain a little mass of matter, believed to be nervous, from which a few filaments proceed. This is the only trace of a nervous system found in the Cœlenterata. Some species of Ctenophora are luminous at night.

Class 2.—Hydrozoa.

188. The **Hydrozoa** (Gr. *hudra*, a water dragon; *zoon*, an animal) *are cœlenterate animals, in which the digestive sac is not separated from the body cavity by any intervening space as in the Actinozoa, the alimentary canal and the somatic cavity forming one continuous tube. The reproductive organs are external, being attached to the outer surface of the body wall.*

189. This class includes the fresh-water hydra, **Sertularidæ** or sea-firs, the **Corynidæ** or club-hydras, the **Lucernaridæ** or lantern-hydras, and the **Medusæ** or jelly-fishes. Some are simple, others form colonies by budding or self-division. They are cellular in structure, and the body consists of two layers, ectoderm and endoderm. The mouth is surrounded by **tentacles**, by means of which the animals seize their prey. They possess **thread-cells**, which are found in the tentacles, as well as in other parts of the body. The mouth leads into a digestive sac, "the outer wall of which is in direct contact with the water in which the animal lives." This sac leads by a wide aperture to the general cavity of the body, which is merely a continuation of the digestive tube. Thus, if a vertical section be made of a hydrozoon, only a single tube will be cut through; whereas, if a similar section be made of an actinozoon, a double tube will have been dissected. "If you could suppose the stomach of a hydrozoon thrust into that part of the body with which it is continuous, so that the walls of the body should rise round it and form a sort of outside case, containing a prolongation of the general cavity, the hydrozoon would be converted into an actinozoon." *(Huxley)*.

190. In the compound Hydrozoa, the individual zooids which form a colony are denominated **polypites,** to distinguish them from the **polypes** in the Actinozoa, and the **polypides** in the Polyzoa. The entire body of a hydrozoon, whether simple or compound, is called the **hydrosoma** (Gr. *hudra*; and *soma*, body); the substance which is common to the various polypites in a colony, the cœnosarc.

The compound Hydrozoa (*sertularia*, &c.,) were formerly placed along with the Polyzoa, both of which were popularly called **Corallines.** These two groups are, however, widely different in internal structure.

191. (1.) The Polyzoa possess an alimentary canal, which is completely shut off from the general cavity of the body, and communicates with the exterior by an anal opening. The alimentary tube, in the Hydrozoa, opens directly into the body cavity, and is not separated from the body walls.

(2.) The Polyzoa have a distinct nervous system. No nerves have been found in any hydrozoon.

(3.) In the Polyzoa the reproductive organs are internal. In the Hydrozoa they form "outward processes of the body wall, and are directly in contact with the surrounding medium."

192. The **fresh-water hydra** may be taken as an example of this class. This animal is common in ponds, and may be found attached to the stems of aquatic plants. In its contracted state it is a little mass of green jelly, about half the size of a pea. If this little gelatinous ball is placed in a glass of water it will manifest its true form. The body is a tube, about half an inch long, and not much

thicker than a sewing thread. It consists of two layers—the outer, the **ectoderm**; the inner, the **endoderm**. Its base is furnished with a sucker by which it is able to attach itself to some solid object. The attachment, however, is not permanent, as it is able to change its position when it wishes to do so. Its mouth is surrounded by a number of tentacles which it has the power of lengthening or shortening with great facility. When completely retracted, they have the appearance of little warts of jelly. The tentacles are hollow tubes, and are formed by prolongations of both the ectoderm and endoderm. They communicate at their bases with the cavity of the body. It is very greedy, and any small animal which it seizes with its tentacles is soon benumbed by the thread cells with which they are furnished. In the brown hydra the tentacles are sometimes eight inches long.

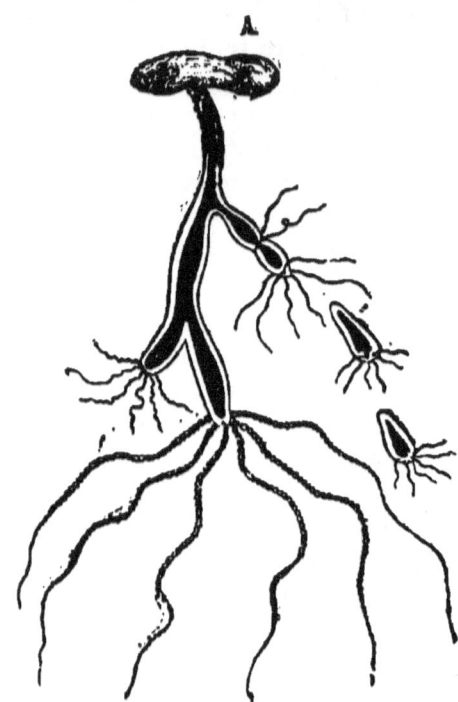

Fig. 56.—Fresh-water Hydra (enlarged).
It is attached to a fragment of duck-weed, and comprises three individuals, of which the largest has the tentacles more expanded than the others. The black lines represent the digestive tube. The individual to the right has been tied by a thread to show how the animals may be multiplied by division. Below are two detached individuals supposed to be formed in this way.

These animals feed on minute worms, insects, Rotifera, water-fleas, and other small Crustacea. The mouth leads into a digestive tube, which opens into the general cavity of the body. This cavity contains water, with which the particles of food are mixed. The indigestible portions of the food are expelled through the mouth.

If a hydra is cut into two pieces, one portion will develop a new trunk and base, the other will soon throw out new tentacles. In this way two perfect animals may be formed from a single individual. If it be cut into half a dozen or more pieces, each piece will develop the wanting parts, and will grow into an animal as perfect as the original. It may even be turned inside out without seeming to suffer any inconvenience.

193. Reproduction, in the hydra, is effected in two ways—by budding, and by ova. During the summer season, little wart-like buds appear on the surface of the body. These increase in size, and a mouth is formed, which is soon surrounded by a circle of tentacles. In its imperfect state, it receives nutriment from the stomach of the parent; but when the tentacles are formed it begins to procure food for itself, and the channel of communication becomes closed up. When it has arrived at maturity, it detaches itself, and becomes independent.

In autumn, ova and sperm cells are formed in little elevations of the outer surface of the body wall. These are deposited simultaneously in the water, and the fertilized ova remain in the bottom of the pond, until the next spring, when they are hatched. Very few hydras, it is believed, live through the winter. The embryo, at first, swims about by the action of the cilia with which it is covered. It soon fixes itself by one extremity, loses its cilia, and obtains a mouth and tentacles.

194. The **Corynidæ** (Gr. *korune*, a club), or **club-hydras,** are marine animals, with the exception of one genus which inhabit fresh water. Some of them are simple, and resemble the fresh-water hydra in structure; but they are permanently attached to a solid object. The greater number are compound, being united by a thread of jelly-like flesh. These are generally provided with a chitinous or horny covering, denominated the **polypary**. This covering does not enclose the polypites in the club-hydras, as it terminates at their bases.

195. The **Tubularia** is an example of this group. It consists of a number of horny tubes which are attached to stones or shells. These tubes are not branched, but are often interlaced with one another. A single polypite is situated at the extremity of each. The tubes are filled with a red jelly-like substance. The polypites are also of a red colour. They are provided with two sets of tentacles, one short row being placed round the mouth, and a longer row near the base.

196. Development.—In the compound Hydrozoa, the buds, instead of being separated from the original stock, as in the fresh-water hydra, remain permanently attached to it, and soon begin to develop other buds. In this way a colony is formed. There are often two different sorts of zöoids. One kind are without reproductive organs, their function being merely to procure food for the colony. These are denominated **alimentary zooids**, and taken collectively, are called the **trophosome** (Gr. *trepho*, I nourish; *soma*, body). These go on increasing by budding; but, at a certain stage in the life of the colony, a second set of buds is produced, furnished with ova and sperm cells. These buds are termed **gonophores** (Gr. *gonos*, offspring; *phero*, I bear). These gonophores are often much larger than the alimentary zöoids, and are of three different kinds.

197. One kind consists of a protuberance formed by the ectoderm and endoderm. They are closed without, but maintain an internal connection with the cœnosarcal fluid. They produce fertile ova.

198. A second kind are bell-shaped discs, which are attached to the polypary by the base, the cavity of the bell being turned outwards. There is a kind of handle within the cavity of this bell, occupying the place of the clapper, and termed the **manubrium** (Lat. *manubrium*, a handle.) From the point where the manubrium is attached to the disc, four canals proceed to the margin of the bell. All these open into a circular canal which surrounds the margin. The ova and sperm cells are produced either in these canals or in the manubrium.

199. The third kind are similar to the preceding, but become detached and lead an independent existence, swimming about by the alternate contraction and expansion of their discs. A mouth is developed at the outer extremity of the manubrium, and the polypite becomes self-feeding. The mouth of the bell is closed, with the exception of a central aperture, by a delicate membrane called the **veil**. A number of tentacles are produced round the margin. As soon as fertile ova are developed, the creature dies.

These bell-shaped gonophores are commonly called sea-jellies or sea-medusæ, and were formerly believed to be distinct animals. The margin of the bell in some groups is adorned with a number of bright spots, which are supposed to be either organs of sight or hearing. These are called the **naked-eyed medusæ**. They vary much in size, some of the smaller kinds being not larger than the head of a pin. Many of them have brilliant colours, and are very beautiful. They are described by the late Professor Edward Forbes as "gorgeous enough to be the diadem of the smallest of the sea-fairies, and sufficiently graceful to be the night-cap of the tiniest and prettiest of mermaids." They are generally phosphorescent, and contribute largely to the luminosity of the sea. The ova developed by the medusæ, instead of reproducing animals of the same kind, grow into fixed plant-like organisms, similar to those from which the gonophores were detached, thus furnishing an example of what has been called, "alternation of generations;" that is, "an individual is not at all like its mother, but exactly resembles its grandmother."

200. The **Sertularidæ** (Lat. *sertum*, a wreath), or sea-firs, are more generally known than any of the compound Hydrozoa. The polypary is much branched, often resembling a minute fern. They may readily be found on the sea shore, sometimes attached to the shells of the mussel and oyster. They are all compound animals, and are confined to the sea.

201. In the **Corynidæ**, the horny polypary merely extends to the base of the polypites; but in the Sertularidæ, it is prolonged into little cups *(Hydrothecæ)*, within which the polypites can withdraw themselves.

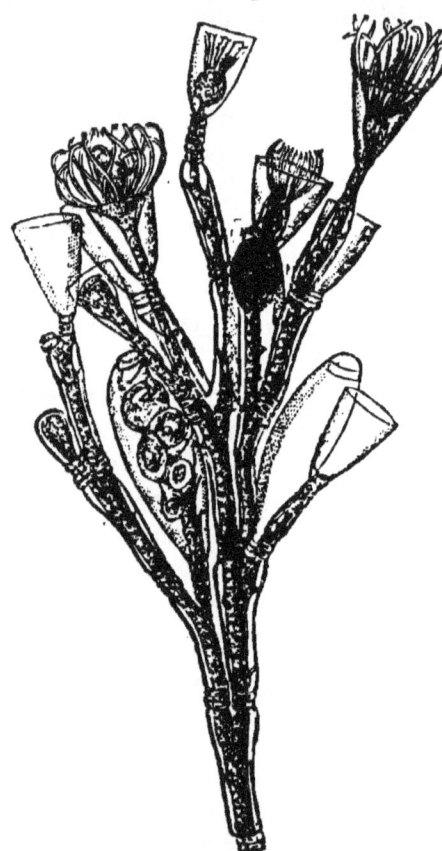

Fig. 57.—CAMPANULARIA (magnified).
Allied to *Sertularia*. Some of the polypites are retracted, others expanded.

The zöoids closely resemble the hydra in structure. At the lower extremity of the body cavity, there is a minute opening which communicates with the hollow tubes of the polypary. These tubes are filled with a jelly-like fluid, which connects all the polypites together into one living mass. To this fluid each individual polypite contributes nutriment, through the aperture at the base of the body cavity. It is kept in constant circulation, probably by the action of cilia.

The ordinary polypites, or **alimentary zooids**, are not provided with reproductive organs. At certain seasons of the year, generative buds are formed, which are much larger than the ordinary cups in which the polypites reside. These buds, being destitute of mouth and tentacles, obtain their nourishment from the cœnosarcal fluid, which permeates the tubes of the polypary. In these cells ova are formed. These, when hatched, become oval-shaped, and swim about for a time by the action of cilia, with which they are covered. At length they attach themselves to some solid object, lose their cilia, and form fresh colonies by budding. In the sertularians, the generative buds are *not* detached, as in the club-hydras.

The extinct Graptolites, found so abundantly in the Silurian rocks, closely resemble the Sertularia. It is believed, however, that they were free-swimming, instead of being permanently attached like the Sertularia.

202. The **Lucernaridæ** (Lat. *lucerna*, a lamp), or lantern-hydras, are represented in our seas by one species, which has an umbrella or cup-shaped body of a gelatinous structure. It attaches itself, by the base, to sea-weeds; but can remove itself at will, like the fresh-water hydra. It is able to swim about, motion being produced by the alternate contraction and expansion of its umbrella. Tufts of short tentacles are placed, at intervals, round the margin of the cup. There is a single polypite in the centre, the mouth of which is surrounded by four lobes. Ova and sperm cells are formed within the body itself, there being no generative buds.

203. A little animal belonging to this group, about half an inch long, denominated the **hydra-tube**, from its resemblance to the fresh-water hydra, forms large colonies by budding. The polypites are usually without reproductive organs. In some cases, the hydra-tube becomes lengthened, and surrounded by a number of grooves. These grooves deepen, and acquire notched margins. In this stage, the polypite has the appearance of a pile of minute cups or saucers. The tentacles which originally surrounded the mouth disappear, and a new set sprout out near the base of the hydra-tube. These

saucers gradually drop off, and appear as free-swimming medusæ. Although originally very small, they soon increase in size, and become, in some cases, of great bulk, measuring seven feet across the disc, with tentacles fifty feet long. These gigantic structures are the **sea-blubbers, sea-nettles,** or **jelly-fish,** so common, in the summer season, round our coasts.

204. These animals are largely composed of sea water.

FIG. 58.—*Aurelia*, one of the jelly-fishes.

Professor Owen has calculated that a jelly-fish, 2 lbs. weight, does not contain more than 30 grains of solid matter. Some of them are furnished with **thread cells,** whose barbed points are able to penetrate the human skin, as many a bather knows to his cost. The principal mass of the body consists of a bell-shaped disc, or umbrella, from the lower part of which a single polypite is suspended. The mouth of this polypite is surrounded by four long processes, or arms. A number of canals (generally eight) proceed from the extremity of the polypite to the margin of the umbrella. Near this margin they are broken up into a great number of tubes, and form a kind of network. All these communicate with a circular canal which runs round the margin of the disc. This margin is furnished with a fringe of tentacles, which are supplied with branches from the circular canal. A number of coloured spots, connected with little sacs filled with fluid, and containing stony particles, are also placed round the margin. These specks are concealed by an expansion of the ectoderm, shaped like a hood. On this account, the name, **hidden-eyed medusæ**, has been applied to this group of jelly-fishes.

205. The following are the principal points of difference between the **hidden-eyed** and **naked-eyed** medusæ:—

In the **naked-eyed medusæ**, the cavity of the disc is furnished with a veil; the radiating canals are usually four in number; these canals never form a network, and the **eyes** are always uncovered.

The **hidden-eyed medusæ** have usually eight canals, which form a network; the **eyes** are covered, and the mouth of the disc is not protected by a veil.

206. The **reproductive** organs "form a conspicuous cross, shining through the thickness of the disc." After swimming about for a time, they develop fertile ova, and then die. The ova, however, instead of producing gigantic sea blubbers, result in minute hydra-tubes, which form colonies as before. Thus we are furnished with another instance of the "alternation of generations,"

Sub-Kingdom VII.—Protozoa.

Class 1.—Infusoria.

207. If a quantity of boiling water be poured upon a few stalks of hay, or other vegetable matter, and allowed to stand for a few days: if a drop of this infusion be then taken and examined by a microscope, it will be found to contain a multitude of minute creatures, moving about with great agility. A like result will be produced by placing the vegetable substance in a glass of water, and exposing it for a few days to the rays of the sun. These animals have been called **Infusoria,** because they were first met with in vegetable infusions. The early observers included the Rotifera in this class; but, as these are much more highly developed animals than the Infusoria, they have now been placed in the sub-kingdom *Annuloida*.

208. The animals produced at first are called **monads,** and are extremely small. "If arranged side by side—in contact with each other like the beads of a necklace—twelve thousand of them could be placed within the length of a single inch." If the infusion be allowed to remain for some time longer, fresh forms will be produced, some of them considerably larger than the monads.

The Infusoria are not confined to infusions of animal or vegetable matter, as they abound in stagnant water everywhere, and are even found in the sea.

They may be defined as *Protozoa, which possess a permanent mouth, and short gullet, which terminates in the central mass of sarcode. Their bodies consist of three distinct layers, the outer one being generally furnished with cilia. They differ from the Rhizopoda, also, in not having the power of protruding pseudopodia.*

209. Some of the Infusoria are free swimming, others are attached to aquatic plants by a slender stalk. **Paramœcium** (Gr. *paramekes,* oblong) may be taken as an example of the unattached groups. It is in shape somewhat like a slipper, and is covered with minute cilia, which are kept constantly in motion. These cilia enable the animal to move rapidly through the water, and also bring a supply of minute particles of food to the mouth. The outer covering of the body is a delicate membrane, called the **cuticle**. Beneath this there is a layer of somewhat firm jelly, which has been termed the **cortical** (Lat. *cortex*, bark) layer. This passes gradually into a semi-fluid mass which occupies the whole interior of the body. No definite structure has been detected in any of these layers.

There is a funnel-shaped mouth, which is succeeded by a short tube, lined with the cuticle. This tube opens directly into the internal sarcode, and conveys such particles of food, mixed with globules of water, as the currents produced by the cilia may have brought. The name **Polygastrica** (Gr. *polus*, many ; *gaster*, stomach), was applied to the Infusoria by Ehrenberg, because he believed these minute globules of water to be so many stomachs. Indigestible particles pass out by an opening situated near the mouth. This anal aperture does not seem to be connected with any tube corresponding to an intestine.

At one or two points there are certain clear spaces in the cortical layer which open and close alternately. These are denominated **contractile vesicles**. When expanded, they appear to be filled with a clear fluid, which is probably water. When contraction takes place, this fluid disappears. Some observers are of opinion that these vesicles communicate with the external medium, and are connected with canals which radiate through the central sarcode. It is probable that this apparatus is of a respiratory nature, bearing some analogy to the water vascular system of the Annuloida.

A little oval mass has been observed in another part of the cortical layer, to which the name **nucleus** has been given. Besides this, there is a smaller round mass which has been called the **nucleolus**. These are believed to be generative organs, the nucleus being an ovary, and the nucleolus a mass of sperm cells. It has, therefore, been concluded that these animals are hermaphrodite, reproducing their kind by fertile ova.

210. Another mode of reproduction is termed **fission** (Lat. *fissus,* a cleft), or self-division. The body of a Paramœcium splits into two parts, each of which becomes an independent being. A well fed animal of this group has been observed to divide in this manner every twenty-four hours.

211. The **Vorticella** (diminutive of Lat. *vortex,* a whirlpool), or bell-animalcule, may be taken as the type of the fixed forms of Infusoria. It may be found attached to the stems of the duck-weed, and other aquatic plants. To the naked eye a group of these animals gives a musty appearance to the stem of the plant on which they are located. When viewed by a microscope they appear as beautiful little bells or vases, attached to stalks which are eight or nine times their own length. This stalk is a tube, and the animal has the power, when alarmed, of contracting it into a spiral form by means of a filament which passes through it in a longitudinal direction.

At the margin of the bell there is a projecting rim which surrounds a circular space called the disc. This disc is separated from the rim by a groove, and is furnished at its outer margin by a fringe of cilia, forming a spiral line which is carried one or more times round the circumference. The animal has the power of retracting the disc with its ciliated fringe into the interior of its body.

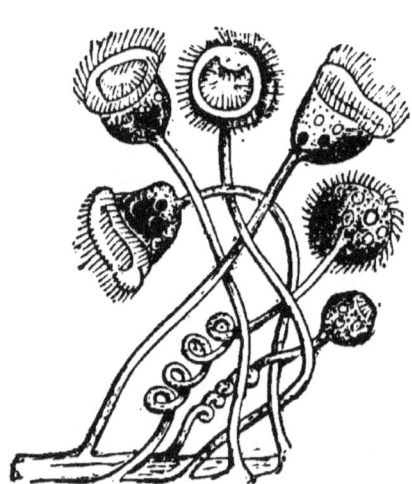

Fig. 59.—Vorticella.

The layers of the body are similar to those found in

Paramœcium. There is an outer **cuticle, a cortical** layer, and a central mass of sarcode. The cortical layer contains a contractile vesicle and a nucleus.

The mouth is situated near the edge of the disc, the fringe of cilia being continued into the œsophagus. The particles of food brought by the cilia, as well as the entire contents of the body, perform a slow rotatory movement through the interior. Indigestible particles are ejected through an anal opening, situated near the mouth.

212. **Development.**—Reproduction in Vorticella is effected by fission, gemmation, and encystation. The first of these modes is thus described by Gosse :—

"One of the full grown bells begins to alter its form, becoming first globular, then a flattened sphere; presently a slight notch or depression is observed in the upper part of the outline, and it soon becomes apparent that this depression is a constriction extending all round, which gradually becomes deeper and more marked. As the two divisions become more separate, each assumes an oval form, united at length to its fellow only at the base. At this time the motion of the cilia is plainly visible, forming a circle in each within the body, near the summit. As the process goes on, the connection between the two is reduced to a mere point, and they become capable of separate motion, so far as to diverge and look in opposite directions; the point of union being the common stem. At this stage we may observe that the bell which is destined to remain is open at the top, within which the ciliary waves are chasing each other in continuous wheels, the other bell being closed at the summit.

"But on the latter, which is ultimately to be thrown off, a new and highly interesting phenomenon appears. The cilia, which before the division had played around the mouth, have become obliterated, probably by absorption; the orifice at that extremity has closed up permanently, for this is to be the base of the new animal; and a new bell-mouth and a new wheel of cilia are to be formed at the opposite end, which at present remains attached to the common stem. The first indication we can detect of this new formation is a very slight movement in the water, a little quivering around what we must yet call the basal part. Presently there appear waved hairs, which seem very flexible, and the motion of which resembles that of a fringe of loose silk moved through the water, an action very different from the regular waves of perfect cilia. These waving hairs increase rapidly in length, and in the vigour and rapidity of their undulations, which gradually become decidedly rotatory, producing at length strong currents in the surrounding water, and imparting a tremulous motion to the whole bell.

"It is evident now that the separation is imminent, for the minute point of connexion cannot long withstand the rushing current of those rotatory paddles. At length the bell suddenly shoots away, gliding with great swiftness through the water, borne by its numerous paddles, and whirls about for a while in a headlong giddy manner. At length it chooses a place of rest, becomes stationary, fixes itself by that end which had formerly been the mouth, but is now closed up, and presently begins to rise by the development of a slender stalk which, though minute at first, increases in length until it attains the original dimensions."

In reproduction by gemmation, the **bell** is not split in two, a bud being produced near its extremity by the expansion of the cortical layer. This bud is at first nourished by the parent animal, but when it obtains its circle of cilia it detaches itself, and swims about for a time, then develops a stalk, and becomes fixed as before.

213. At certain seasons the Vorticella coats itself with a **cyst** (Gr. *kustis*, a bladder) of gelatinous matter, the cilia and stalk disappearing. After a time the **cyst** bursts, and a number of germs are set free, which, after swimming about for a time, develop stalks and become bell-animalcules like the original. This is denominated reproduction by **encystation**; whether it is sexual or otherwise, has not been ascertained.

Class 2.—Rhizopoda.

214. The **Rhizopoda** (Gr. *rhiza*, a root; *poda*, feet) may be defined as *sarcode animals destitute of a mouth, and having the power of sending out pseudopodia.*

To this class belong the *Amœba*, the *Foraminifera*, the *Radiolaria*, and *Spongida*, each of which will require a separate description.

215. (1.) The **Amœba** (Gr. *amoibos*, changing), or Proteus animalcule, is a microscopic animal commonly met with in stagnant ponds. It is the simplest of all animals. It consists of a little mass of jelly-like sarcode, resembling the white of an egg, or rather a drop of gum water. It is destitute of all those organs which are usually considered to be essential to life. There is no mouth or digestive apparatus, no heart, no circulatory organs, no respiratory organs, no nervous system. The animal is

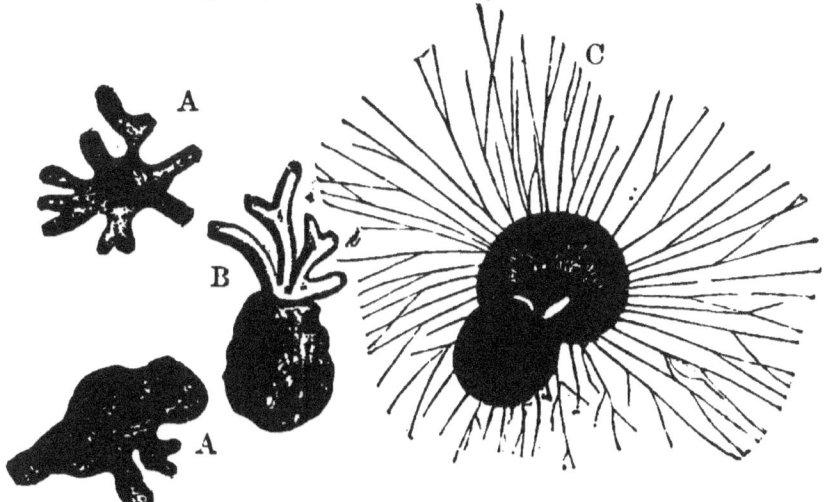

Fig. 60.—AA, Different forms of *Amœba*; B, *Amœba diflugia*, having the pseudopodia protruded from one extremity only; C, *Milliola*, one of the *Foraminifera*, in which the pseudopodia are protruded only from the mouth of the shell.

able to use any part of its body as a hand or a foot, and to extemporize a mouth and alimentary canal anywhere that may best suit its exigencies at the time.

It constantly alters its shape, pushing out from various parts of its substance blunt, finger-like processes, denominated **pseudopodia** (Gr. *pseudos*, false; *poda*, feet). These pseudopodia procure food for the animal, and enable it to change its position slowly. When a pseudopodium comes in contact with some particle of animal or vegetable matter, it surrounds it with its jelly. The rest of the sarcode flowing up to this point, the food is inclosed in the substance of the body, where it is digested. It soon pushes out another process in a different direction, when a similar result follows. Sometimes six or eight pseudopodia are sent out at the same time. In this way food is procured, and locomotion effected simultaneously. As it is destitute of excretory organs, indigestible particles are pushed out through the sarcode.

The outer layer of the body, termed the **ectosarc** (Gr. *ecto*, without; *sarx*, flesh), is a little more solid than the interior, which is called the **endosarc** (Gr. *endo*, within).

Each particle of food carries with it a little globule of water. These globules are visible in the interior of the animal, and were formerly believed to be stomachs. A **contractile vesicle,** similar to that observed in the Infusoria, has been found in the Amœba. It is filled with some clear fluid, and contracts and expands at pretty regular intervals. There is also a **nucleus,** which some suppose to be connected with reproduction: but there is no certain knowledge on this point.

The most usual mode of reproduction which has been observed is by **fission,** or self-division. The animal in certain cases splits into two parts, each of which becomes a distinct individual. Occasionally, a detached pseudopodium is developed into a separate animal.

216. (2.) The **Foraminifera** (Lat. *foramen*, a hole; *fero*, I bear) differ from Amœba in being covered with a shell which is usually calcareous, although in rare cases it is composed of particles of sand united together. These shells assume a great variety of forms, and many of them are extremely beautiful. They are so exceedingly small that "hundreds of them would hardly weigh a grain." In some instances, the shell consists of only a single chamber; but usually there are a number of cells communicating with one another by minute openings. Some of them are whorled like the shell of the nautilus, and, on this account, were formerly believed to be minute Mollusca.

The animal which secretes those beautiful shells from the sea-water seems to be of even a lower organization than the Amœba, as it does not contain any **nucleus** or **contractile vesicle**. It is very extraordinary that such a lowly creature should be able to execute such a beautiful piece of workmanship. We have here an illustration of the fact, that "life is the *cause* of organization, and not the *consequence* of it."

A foraminifer consists originally of a little mass of jelly of a red or yellow colour. The outer coating of this jelly secretes a shell consisting of a single chamber. As the animal continues growing, it usually throws off buds, each of which forms a fresh chamber. Thus, one of these shells, minute as they are, generally contains a small colony of animals all united together into a living mass by the canals which connect the chambers. In one of the commonest groups (*Globigerina*), "when the process of budding has produced a series of sixteen segments, the next bud detaches itself, and begins to form a separate shell." (*Dr. Carpenter.*)

The pseudopodia, instead of being blunt processes, as in Amœba, are long threads of inconceivable fineness. These threads interlace one another, forming what has been termed a sort of "animated spider's web." The interior jelly is constantly sending out fresh threads in search of particles of food, and receiving those which are drawn back. Sometimes the pseudopodia are emitted through the mouth of the shell, but more generally through minute apertures (*foramina*) in the walls of the chambers. On this account the name *Foraminifera* has been given to the whole group. Strictly speaking, the pseudopodia are

150   ZOOLOGY.

sent out from a film of jelly which surrounds the surface of the shell. This, however, is in direct communication with the sarcode in the interior.

These Foraminifera are extremely abundant, being,

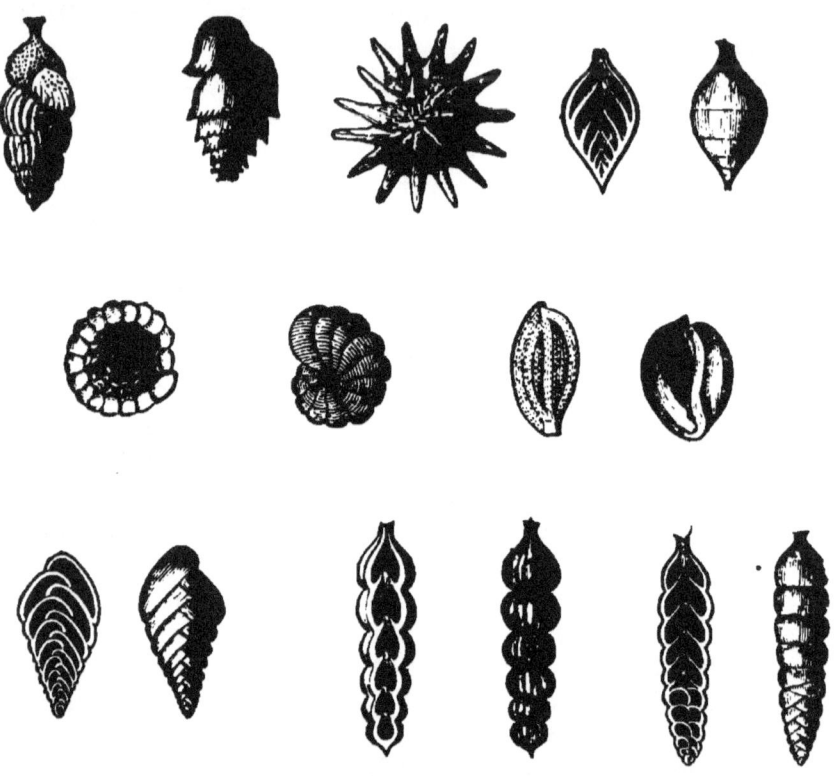

Fig. 61.—Group of Foraminifera.

probably, more numerous than all other animals together. They are generally marine, and, in some countries, form a large portion of the sand on the sea-shore. Along a part of the west coast of Ireland, the whole sea-beach for miles is mainly composed of their remains. They also cover a wide area of the bottom of the North Atlantic. They been have dredged up from a depth of three miles in this ocean. A mass of these shells has the appearance of white mud. Chalk, which forms in many countries an important portion of the earth's crust, is now known to be almost wholly composed of a single species of Foraminifera (*Globigerina*). Several other limestones also have been mainly formed by them. The earliest vestige of animal life found in the rocks of the earth—the **eozoon** of the Lawrentian strata—is a foraminifer.

The modern forms are all microscopic, but the extinct **nummulite** (Lat. *nummus*, money ; Gr. *lithos*, a stone) which has formed large masses of limestone, extending from France to China, is sometimes three inches in circumference. The eozoon seems also, by continuous budding, to have attained a large size.

217. (3.) The **Radiolaria** (Lat. *radius*, a ray) form the third group of Rhizopoda. They are so called because the pseudopodia surround the body like rays. They are distinguished from the Foraminifera by having a **siliceous** instead of a calcareous skeleton, which sometimes takes the form of a shell; and, in other cases, consists of loose **spicula** (Lat. *spiculum*, a point), or needles. The best known members of this group are the **Polycystina** (Gr. *polus*, many; *kustis*, a bladder), which are even smaller than the Foraminifera. They are furnished with minute glassy shells, which, under the microscope, are objects of extreme beauty. The animal consists of a little mass of brown sarcode, which does not completely fill the shell. The pseudopodia are protruded through apertures in the shell, but they do not form a network. The white mud dredged up from the bottom of the Atlantic contains a considerable proportion of these shells, mixed with those of the Foraminifera. They are also found very abundantly as fossils in some tertiary rocks.

218. The **Thalassicollida** (Gr. *thalassa*, the sea ; *kolla*, glue) are gelatinous animals found floating near the surface of the ocean. They are sometimes an inch in diameter. Their bodies are surrounded by radiating pseudopodia, which sometimes form a network. The skeleton consists of a series of flinty needles scattered through the sarcode.

219. The **Acanthometra** (Gr. *acanthos*, a thorn; *metra*, the womb) have a skeleton of flinty spines which radiate from the centre of the gelatinous body. Some of the pseudopodia pass through canals in the interior of the spines; others proceed directly from the mass of sarcode. They do not form a network. They are smaller than the Thalassicollida, and are found floating at the surface of the sea.

220. The **Actinophrys**, or sun-animalcule, is also referred by Dr. Carpenter to this group.

221. (4.) **Spongida.**—Until recently, the sponges were considered to be vegetables; but they are now known to be entitled to a place in the animal kingdom. What we call a sponge is merely the framework or skeleton formed by a mass of animals, each of which resembles an Amœba, and contains a nucleus. These skeletons are generally horny, as in the sponge of commerce; but there are also calcareous and siliceous sponges. The horny sponges are often strengthened by spicula, or needles of siliceous or calcareous matter. These skeletons are invested by a jelly-like mass, which resembles in appearance the white of an egg. When the sponge is lifted out of the water, this jelly drains off. This gelatinous substance is permeated in every direction by canals which communicate with the surface by a great number of small openings, and a few large ones. Corresponding openings may be observed in the fibrous skeleton. Currents of water enter by the smaller or **inhalant apertures**; and, after passing through the system of canals, are driven out through the larger or **exhalant apertures**. Careful examination has shown that these currents are produced by cilia, which are situated in spherical dilatations of the canals in the inner layer of the sponge. These cavities are surrounded by particles of sarcode, "each of which is furnished with a vibratile cilium." The cilia all work so as to drive the water out by the exhalant apertures. The currents thus produced bring such particles of food as may be floating in the water, and

PROTOZOA—RHIZOPODA. 153

these are assimilated by the sponge particles in a similar

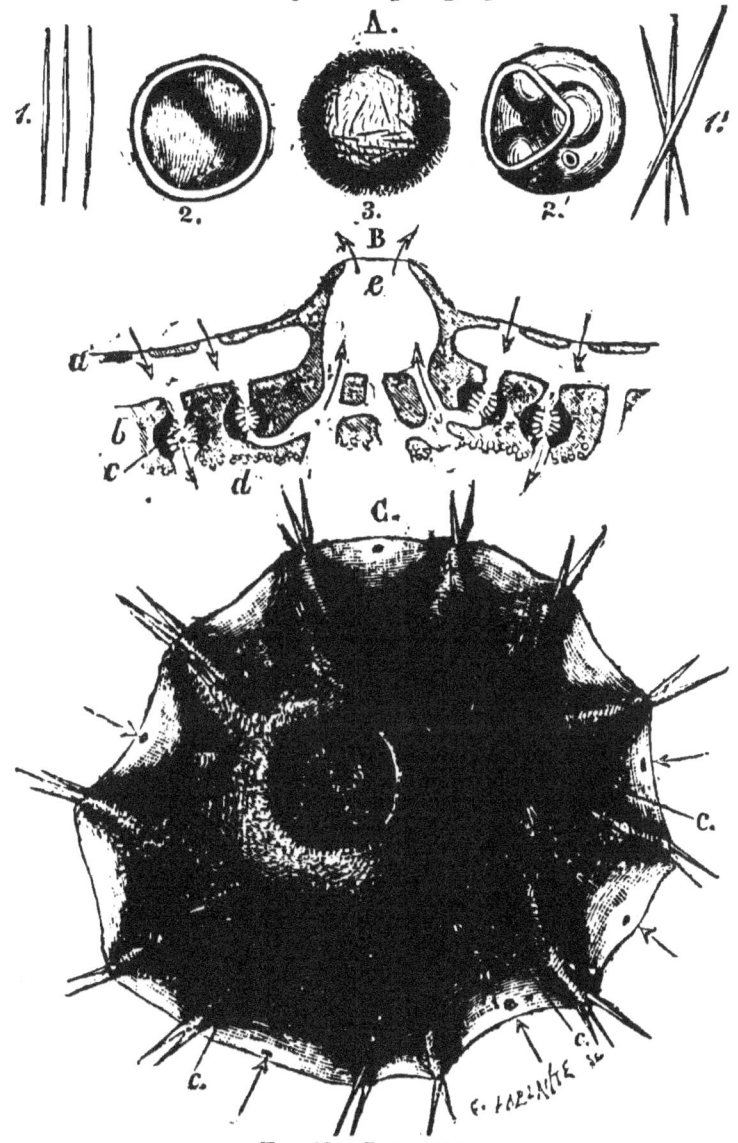

FIG. 62.—SPONGILLA.

A—1, 1′, *spicula*; 2, 2′, gemmules; 3, ciliated embryo. B, section of *spongilla*—the arrows indicate the direction of the currents; *a*, outer layer; *b*, inner layer; *c*, ciliated chamber; *e*, exhalant orifice. C, Mass of *spongilla*—the arrows indicate the inhalant apertures; *c c*, openings leading to ciliated chambers; single exhalant aperture in the centre.

manner to that by which the Amœba is fed. "The sponge represents a kind of subaqueous city, where the people are arranged about the streets and roads in such a manner that each can easily appropriate his food from the water as it passes along." *(Huxley.)*

222. Reproduction in the sponges is effected in two different ways. In the fresh-water sponge (*spongilla*), towards the approach of winter, extremely minute bodies called **seeds,** or **gemmules,** are formed in the inner layer. One of these gemmules consists of a number of sponge particles, coated over with **a cyst.** This cyst has imbedded in it spicula of a very curious form, consisting of an axle with a toothed wheel at each end. During the winter the spongilla dies; and, on the arrival of spring, the sponge particles pass out through an opening in the cyst, and become sponges like the original.

During the summer months some of the sponge particles are changed into ova, others into sperm cells. The resulting embryos are provided with cilia. After swimming about for a time they lose their cilia, attach themselves to some solid object, and are developed into sponges.

223. Distribution.—Sponges are found both in salt and fresh water, and are met with in most parts of the world. The sponges of commerce are all marine, and are confined to the warmer regions. The best sponges are obtained from the Greek islands, a coarser kind being procured from the Bahamas. Fossil sponges are common in various geological formations, but are especially abundant in the chalk. The flint nodules which abound in this formation are believed to have been mainly formed by them. It is an interesting and suggestive fact that siliceous sponges, similar to those found in the chalk, have been recently dredged from the bottom of the Atlantic, along with the shells of the Foraminifera.

Class 3.—Gregarinida.

224. The **Gregarinida** (Lat. *grex*, a flock) are internal parasites, infesting the alimentary canal of the cockroach, the earthworm, and other animals. They are destitute of a mouth, and do not protrude pseudopodia. Like other internal parasites, they subsist by imbibing the juices of the animals in which they live. They have the appearance of minute worms, and are sometimes about half an inch long, though generally much smaller. The outer coating is membranous, the interior being filled with sarcode. At one point in the central jelly there is a vesicle, or **nucleus**, enclosing a solid particle or **nucleolus**. These are probably reproductive organs.

225. In certain cases these animals become changed into minute balls, enclosed by a **cyst**. The nucleus and nucleolus then disappear, and the sarcode is formed into little masses, which are at first round, but afterwards lengthen out, and become pointed at each end. These are called **Pseudo-Navicellæ** (Navicella being a microscopic plant or *diatom*). The cyst afterwards bursts, and the Pseudo-Navicellæ pass out of the body of the animal in which the Gregarinæ reside. Some of these are taken into the body of another animal—an earthworm, for example; and then the little mass of sarcode, inclosed within the shell of the Pseudo-Navicella, is set free, and begins to thrust out processes after the manner of an Amœba. Ultimately, this little animal is developed into a Gregarina like that from which it proceeded.

QUESTIONS—V.

1. Give the derivation of the term *Actinozoa*.
2. Give examples of this class.
3. What are its principal characteristics?
4. Explain the terms MESENTERIES, ECTODERM, ENDODERM, CŒNOSARC, ACTINOSOMA.
5. Give an account of the structure of the SEA-ANEMONE.
6. What is coral?
7. What is the distinction between SCLERODERMIC and SCLEROBASIC corals?
8. Give an account of the various forms of CORAL REEFS.
9. What is an ATOLL?
10. How does the red coral of the Mediterranean differ from the reef-building corals?
11. Describe the *Ctenophora?*
12. Define the class *Hydrozoa?*
13. Give examples of this class.
14. How does the section of a *Hydrozoön* differ from that of an *Actinozoön?*
15. Distinguish between POLYPES, POLYPITES, and POLYPIDES.
16. What is the *hydrosoma?*
17. What are the leading distinctions between the *Hydrozoa* and the *Polyzoa?*
18. Give an account of the structure and habits of the FRESH-WATER HYDRA.
19. What are THREAD-CELLS?
20. Give an account of the reproduction of the Hydra.
21. Describe the CORYNIDÆ or CLUB-HYDRAS.
22. What is meant by ALIMENTARY ZÖOIDS?
23. What are GONOPHORES?
24. Give an account of the different kinds.
25. What is the MANUBRIUM?
26. Give an account of the development of the NAKED-EYED MEDUSÆ.
27. What is meant by "alternation of generations?"
28. Describe the *Sertularidæ*.
29. How do they differ from the *Corynidæ?*
30. Give an account of the *Lucernaridæ*.
31. What is a HYDRA-TUBE?
32. Give an account of the development of the great JELLY-FISHES.
33. What are the principal points of difference between the HIDDEN-EYED and NAKED-EYED MEDUSÆ?
34. Why are the *Infusoria* so called?
35. Define this class.
36. Give examples.

37. Describe *Paramœcium*.
38. Explain the terms CORTICAL LAYER, CONTRACTILE VESICLE, NUCLEUS, NUCLEOLUS.
39. What various modes of reproduction have been observed in this class?
40. Describe *Verticella*, and give an account of the various modes of reproduction observed in this group.
41. Define the *Rhizopoda*.
42. Give examples.
43. What are PSEUDOPODIA?
44. What is the mode of reproduction in the *Amœba*? Give a description of this animal.
45. What is the origin of the term *Foraminifera*?
46. How do they differ from *Amœba*?
47. Where are they usually found?
48. Describe the shells secreted by these animals.
49. How do the *Radiolaria* differ from the *Foraminifera*?
50. What are the *Polycistinœ*?
51. Describe the *Thalassicollida*.
52. Give an account of the *Acanthometra*.
53. What is the structure of a sponge?
54. What is the nature of the frame-work or skeleton?
55. What are the INHALANT and EXHALANT apertures?
56. How is a sponge fed?
57. Describe the mode of reproduction in the fresh-water sponge.
58. What are SPICULA? What are GEMMULES?
59. What are the *Gregarinidœ*?
60. What is their usual habitat?
61. What are PSEUDO-NAVICELLÆ?

GLOSSARY

Abdomen (Lat. *abdo*, I conceal)—In *Mammalia*, that part of the body cavity which is separated from the thorax or chest by the diaphragm.

Acalephæ (Gr. *akalephe*, a nettle)—Jelly fishes or sea-nettles, so called on account of their ability to sting by means of "thread cells."

Acanthocephala (Gr. *acanthos*, a thorn; *kephale*, the head)—A group of internal parasites, so called because the head is armed with spines.

Acephalous (Gr. *a*, without; *kephale*, the head)—Without a distinct head. The *Lamellibranchiata* are sometimes called *Acephala.*

Actinosoma (Gr *aktis*, a ray; *soma*, body)—A term applied to the whole body of an actinozoon, whether simple or compound.

Actinozoa (Gr. *aktis*, a ray; *zoon*, an animal)—A class of *cœlenterata*.

Albumen (Lat. *albus*, white)—A substance resembling the white of an egg.

Alimentary Canal (Lat. *alo*, I nourish)—The tube through which the food passes.

Allantois (Gr. *allas*, a sausage)—A fœtal membrane in mammals, birds, and reptiles.

Ambulacra (Lat. *ambulacrum*, a garden walk)—The spaces in the tests of sea-urchins and star-fishes, containing the apertures through which the "tube feet" are protruded.

Ametabolic (Gr. *a*, without; *metabole*, change)—A term applied to those insects which do not undergo metamorphosis.

Ammonites (so called from their resemblance to the horns on the statues of Jupiter-Ammon)—A group of extinct chambered shells belonging to the class *Cephalopoda*.

Amnion (Gr. *amnos*, a lamb)—A fœtal membrane in mammals, birds, and reptiles.

Amœba (Gr. *amoibe*, a change)—One of the *Rhizopoda*, which constantly changes its shape.

Amphibia (Gr. *amphi*, both; *bios*, life)—A class of *Vertebrata* adapted for breathing in water when young, and in air when mature. The term was formerly applied to aquatic mammals, such as the seals.

Amphicœlus (Gr. *amphi*, both; *koilos*, hollow)—A term applied to double-concave *vertebræ*.

Amphioxus (Gr. *amphi*, both; *oxus*, sharp)—A little fish (the lancelet) which tapers at both ends.

Anarthropoda (Gr. *a*, without; *arthros*, a joint; *pous*, a foot)—A division of *Annulosa*, so called because they are destitute of jointed limbs.

Anchylosis (Gr. *ankulos*, crooked)—The union of two bones which are immovably united together.

Animalcule (diminutive of animal)—The popular name for a microscopic animal.

Annelida (Lat. *annulos*, a ring)—A class of *Anarthropoda*, embracing the earth-worm and the leech, &c.

Annuloida (Lat. *annulus*, a ring)—A sub-kingdom, including the *Echinodermata* and *Scolecida*.

Annulosa (Lat. *annulus*, a ring)—A sub-kingdom, composed of the *Arthropoda* and *Anarthropoda*. Formerly called *Articulata*.

Antennæ (Lat. *antenna*, the yard of a ship)—The jointed appendages of the head in *Insecta*, *Crustacea*, and *Myriapoda*.

Antennules (diminutive of antennæ)—The shorter pair of antennæ in the *Crustacea*.

Arachnida (Gr. *arachne*, a spider)—A class of *Arthropoda*, including spiders, scorpions, and mites.

Archæopteryx (Gr. *archaios*, ancient; *pteryx*, a wing)—An extinct bird found in the oolitic rocks; the only representative of the order *Saururæ* of Huxley.

Ascidioida (Gr. *askos*, a bag)—A class of *Molluscoida*, so called because they resemble in shape a two-necked bottle. This class is also termed *Tunicata*.

Atlas (Gr. *Atlas*, the god that holds up the earth)—The *vertebra* which supports the skull.

Atrium (a hall)—Applied to the cavity in the *Tunicata* into which the intestine opens.

Auricle (diminutive of *auris*, an ear)—One of the cavities of the heart which receives blood, and transmits it to the ventricle.

Autophagi (Gr. *autos*, self; *phago*, I eat)—A term applied to those birds which are able to run about, and obtain food as soon as they leave the shell.

Aves (Lat. *avis*, a bird)—One of the classes of *Vertebrata*.

Avicularium (Lat. *avicula*, a little bird)—An appendage of the *Polyzoa*, resembling the head of a bird.

Belemnites (Gr. *belemnos*, a dart)—An extinct group of *Cephalopoda*, allied to the cuttle-fishes.

Bilateral (Lat. *bis*, twice; *latus*, a side)—Having two similar sides.

Bivalve (Lat. *bis*, twice; *valvæ*, folding doors)—Applied to the shells of the *Lamellibranchiata* and *Brachiopoda*, which are composed of two pieces.

Brachiopoda (Gr. *brachion*, an arm; *pous*, a foot)—A class of *Molluscoida*, having long ciliated arms, and bivalve shells.

Branchiæ (Gr. *branchiæ*, gills)—The breathing organs of fish and other aquatic animals.

Branchiogasteropoda (Gr. *branchiæ*, gills; *gaster*, the belly; *pous*, a foot)—A group of *Gasteropoda* that breathe by means of gills.

Bronchi (Gr. *bronchos*, the windpipe)—The branches of the windpipe that permeate the lungs.

Bryozoa (Gr. *bruon*, moss; *zoon*, an animal)—Another name for the *Polyzoa*.

Byssus (Gr. *bussos*, flax)—Silky threads by which the mussel and other *lamellibranchs* attach themselves to rocks.

Cæca—Blind processes in the alimentary canal.

Canine Teeth (Lat. *canis*, a dog)—The eye teeth, so called because they are well developed in the dog and other carnivorous animals.

Carapace—The upper shell of crabs and lobsters; the upper half of the case in which tortoises and turtles are enclosed.

Carpus (Gr. *carpos*, the wrist)—The bones that are placed between the fore-arm and the hand.

Caudal (Lat. *cauda*, the tail)—Belonging to the tail.

Cephalopoda (Gr. *kephale*, the head; *pous*, foot)—A class of *Mollusca*, to which the cuttle-fish and nautilus belong. They are so called because the organs of locomotion are arranged round the head.

Cephalothorax (Gr. *kephale*, the head; *thorax*, breast)—That part of the body in *Crustacea* and *Arachnida* which is formed by the union of the head and thorax.

Cervical (Lat. *cervix*, the neck)—Belonging to the neck.

Chætognatha (Gr. *chaite*, hair; *gnathos*, a jaw)—A class of *Anarthropoda*, containing the *Sagitta* only.

Chitine (Gr. *chiton*, a coat)—The horny substance which forms the exoskeleton of insects, &c.

Chrysalis (Gr. *chrusos*, gold)—The pupa state of an insect, so called because it is sometimes of a golden colour.

Chyle (Gr. *chulos*, juice)—The milky fluid which results from the digestion of food.

Chyme (Gr. *chumos*, juice)—The pulpy mass into which food is formed by the action of the gastric juice.

Cilia (Lat. *cilium*, an eye-lash)—Microscopic hair-like organs which, by their movements, produce currents in the water, and thus supply the *Infusoria* and other animals with minute particles of food. They also serve as means of locomotion.

Cirri (Lat. *cirrus*, a curl of hair)—A term applied to the feet of barnacles and acorn-shells, from their resemblance to tendrils. The filaments attached to the arms of the *Brachiopoda*.

Cirripedia (Lat. *cirrus*, a curl; *pes*, a foot)—A division of *Crustacea*, including the barnacles and acorn-shells. They are so called on account of their curled feet.

Clavicle (Lat. *clavicula*, a little key)—The collar-bone of the *Vertebrata*.

Cloaca (Lat. a sewer)—The cavity into which the intestine, urinary, and genital organs open in birds, reptiles, and one order of mammals *(Monotremata)*. An analogous cavity is found in some *invertebrate* animals.

Cocoon (Fr. *cocon*)—The covering of an insect in the pupa state, sometimes consisting of silky hairs.

Cœlenterata (Gr. *koilos*, hollow; *enteron*, an intestine)—The sub-kingdom which includes the *Hydrozoa* and *Actinozoa*. They are distinguished from the *Protozoa*, mainly, in having a *hollow digestive cavity*.

Cœnosarc (Gr. *koinos*, common; *sarx*, flesh)—The common connecting stem in the compound *Hydrozoa*.

Commissures (Lat. *committo*, I join together)—The nerve filaments which unite the ganglia or nerve-centres.

Conchifera (Lat. *concha*, a shell; *fero*, I bear)—A name sometimes applied to the *Lamellibranchiata*.

Condyle (Gr. *kondulos*, a knuckle)—The articulating surface of a bone, especially applied to the articulating surface or surfaces of the occipital bone, which join the skull to the atlas, or first joint of the vertebral column.

Coracoid (Gr. *korax*, a crow)—A second clavicle found in birds, reptiles, and Monotremata. In man, the "coracoid process" of the scapula resembles the beak of a *crow*.

Corallum (Lat. for coral)—The calcareous structure secreted by the tissues of the *Actinozoa*, generally called "coral."

Coriaceous (Lat. *corium*, a hide)—Resembling leather.

Corpus Callosum (Lat. "firm body")—The band of nerve-fibres which unites the two hemispheres of the *cerebrum* in Mammalia.

Corpuscles (Lat. *corpusculum*, a little body)—The rounded solid particles found floating in blood. In most mammals they have the form of "biconcave discs."

Cortical Layer (Lat. *cortex*, bark)—In the *Infusoria* this term is applied to the middle layer of the body. It lines the cuticle, and surrounds the central mass of sarcode.

Cranium (Gr. *kranion*, the skull)—The bony or cartilaginous case which surrounds the brain.

Crop—An expansion of the œsophagus found in birds.

GLOSSARY.

Crustacea (Lat. *crusta*, a crust)—A class of *Annulosa*, including crabs, lobsters, &c. They are so called because they are covered with a shell or hard crust.

Ctenoid (Gr. *kteis*, a comb; *eidos*, form)—A term applied to the scales of fishes which have comb-like margins.

Ctenophora (Gr. *kteis*, a comb; *phero*, I bear)—A group of *Actinozoa*, whose swimming organs consist of comb-like bands of cilia.

Cursores (Lat. *curro*, I run)—An order of birds, including the ostrich, emeu, cassowary, and rhea, which are destitute of the power of flight.

Cuticle (Lat. *cutis*, the skin)—The outer layer of the skin.

Cycloid (Gr. *kuklos*, a circle; *eidos*, form)—A term applied to circular fish scales.

Cyst (Gr. *kustis*, a bladder)—A bladder-like sac.

Dermal (Gr. *derma*, skin)—Belonging to the skin.

Diaphragm (Gr. *diaphragma*, a partition)—The muscle which separates the thorax from the abdomen in the *Mammalia*.

Diatomaceæ (Gr. *diatemno*, I cut asunder)—A group of microscopic plants, provided with siliceous coverings.

Dibranchiata (Gr. *dis*, twice; *branchiæ*, gills)—The cuttle-fish tribe, so called because they possess two gills.

Didelphia (*dis*, twice; *delphus*, the womb)—The marsupial mammals.

Digit (Lat. *digitus*, a finger)—A finger or toe.

Dorsal (Lat. *dorsum*, the back)—Belonging to the back.

Echinodermata (Gr. *echinos*, a hedgehog; *derma*, skin)—A class of animals, including the sea-urchins and star-fishes. It is so called on account of the spines with which the integument is generally furnished.

Ectocyst (Gr. *ektos*, without; *kustis*, a bladder)—The outer covering of the *Polyzoa*.

Ectosarc (Gr. *ektos*, without; *sarx*, flesh)—The outer layer of sarcode in the *Amœba*.

Elytra (Gr. *elytron*, a sheath)—The horny front pair of wings in beetles, which form a case for the protection of the membranous pair.

Embryo (Gr. *en*, in; *bruo*, I swell)—The earliest stage in which an animal may be discerned in the egg.

Encrinite (Gr. *krinon*, a lily)—The "stone lily," a group of *Echinodermata*.

Encystment (Gr. *kustis*, a bladder)—A change undergone by some of the Protozoa, in which they become motionless, and cover themselves with a "cyst" or sac.

Endocyst (Gr. *endon*, within; *kustis*)—The inner layer of the *Polyzoa*.

Endoderm (Gr. *endon;* *derma,* skin)—The inner layer of the *Cœlenterata.*

Endosarc (Gr. *endon;* *sarx,* flesh)—The inner sarcode of the *Amœba.*

Entomostraca (Gr. *entoma,* insects; *ostrakon,* a shell)—A group of minute fresh-water *Crustacea.*

Entozoa (Gr. *entos,* within; *zoon,* an animal)—Intestinal worms.

Epidermis (Gr. *epi,* upon; *derma,* the skin)—The cuticle or outer layer of the skin.

Equilateral (Lat. *æquus,* equal; *latus,* a side)—Equal-sided; applied to the shells of the *Brachiopoda.*

Equivalve—Having equal valves, as in most of the shells of the *Lamellibranchiata.*

Exoskeleton (Gr. *exo,* without)—The outer skeleton of insects, &c., formed by a hardening of the integument.

Fauna (Lat. *fauni,* rural gods)—The native animals of a country or district.

Femur—The thigh bone, placed between the *pelvis* and the *tibia* in the higher *Vertebrates.*

Fibula (Lat. a brooch)—The smaller bone of the leg, corresponding to the *ulna* of the fore-arm.

Fission (Lat. *findo,* I cleave)—A method of reproduction by self-division, found in the *Infusoria* and other animals.

Flora (Lat. the goddess of flowers)—The plants of a country or district.

Foot-Jaws—Certain limbs of the *Crustacea* which serve as masticating organs.

Foot-Tubercles—The unjointed limbs of the *Annelida.*

Foraminifera (Lat. *foramen,* a hole; *fero,* I bear)—A group of *Rhizopoda* which possess minute, calcareous shells. These shells are perforated by minute apertures, through which pseudopodia are protruded.

Furculum (Lat. *furca,* a fork)—The V-shaped bone in birds, formed by the union of the *clavicles.* It is commonly called the "merry-thought."

Ganglion (Gr. a knot)—A mass or centre of nervous matter, containing nerve cells, from which nerve fibres proceed.

Ganoid (Gr. *ganos,* splendour)—A term applied to fish scales, which are composed of an inner layer of bone, and an outer layer of shining enamel.

Gasteropoda (Gr. *gaster,* the belly; *pous,* foot)—A class of *Mollusca* usually furnished with univalve shells. They are so called because the locomotive organ consists of a broad muscular disc, occupying the lower surface of the body.

GLOSSARY.

Gemmation (Lat. *gemma*, a bud)—A method of reproduction by which new individuals are formed from buds which arise on the body of the parent.

Gemmule (Lat. a little bud)—A minute mass of spongy particles covered with a *cyst*.

Gizzard—A part of the stomach of birds, insects, &c., furnished with muscular walls.

Globigerina (Lat. *globus*, a ball; *gero*, I carry)—A group of *Foraminifera*, found abundantly in chalk.

Graptolites (Gr. *grapho*, I write; *lithos*, a stone)—An extinct group of *Hydrozoa*, whose remains are found in the Silurian strata.

Gregarinida (Lat. *grex*, a flock)—A class of *Protozoa*.

Guard—The fibrous sheath which protects the phragmacone of the *Belemnite*.

Hæmal (Gr. *haima*, blood)—Belonging to the blood.

Hectocotylus (Gr. *hekaton*, a hundred; *kotulos*, a cap)—One of the arms of the male cuttle-fishes modified into a reproductive organ.

Hemimetabolic (Gr. *hemi*, half; *metabole*, change)—Insects which undergo incomplete metamorphosis.

Hermaphrodite (Gr. *Hermes*, Mercury; *Aphrodite*, Venus)—Having the sexes combined in the same individual.

Heterocercal (Gr. *heteros*, different; *kerkos*, a tail)—Applied to the caudal or tail fin of fishes when the two lobes are unequal.

Heterophagi (Gr. *heteros*, different; *phago*, I eat)—A term applied to birds which come from the egg in a helpless state, requiring to be fed for a time by the parents.

Holometabolic (Gr. *holos*, whole; *metabole*, change)—A term applied to insects whose metamorphosis is complete.

Holostomata (Gr. *holos*, whole; *stoma*, mouth)—A group of *Gasteropoda*, in which the aperture of the shell is not furnished with a notch or canal.

Homocercal (Gr. *homos*, like; *kerkos*, tail)—A term applied to the tail fins of fishes which are divided into two equal lobes.

Humerus—The bone which connects the radius with the pectoral arch in vertebrates.

Hydrosoma (Gr. *hudra*, a water dragon; *soma*, body)—The entire body of a hydrozoon, whether simple or compound.

Hydrothecæ (Gr. *hudra*; *theka*, a case)—The horny cups inhabited by the sertularian polypites.

Hydrozoa (Gr. *hudra*; *zoon*, an animal)—A class of *Cœlenterata*, exemplified by the fresh-water *hydra* and jelly-fishes.

Hyoid—The tongue bone, so called because in man it is shaped like the letter U.

Ichthyopsida (Gr. *ichthus*, a fish; *opsis*, appearance)—The division of the *Vertebrata* which includes the *Amphibia* and fishes.

Ichthyosaurus (Gr. *ichthus; saurus*, a lizard)—An extinct reptile found in the secondary rocks.

Imago (Lat. an image)—The perfect form of insects.

Incisors (Lat. *incido*, I cut)—The front or cutting teeth of mammals.

Incubation (Lat. *incubo*, I sit)—The act of sitting on eggs, in order to develop the contained embryos.

Inequilateral (Lat. *in*, not; *æquus*, equal; *latus*, a side)—Applied to the shells of the *Lamellibranchiata*, which have the two sides unequal.

Inequivalve—Having two unequal valves.

Infusoria (Lat. *in*, on; *fundo*, I pour)—A class of *Protozoa*, so called because they abound in infusions of animal or vegetable substances.

Insecta (Lat. *inseco*, I cut into)—A class of *Arthropoda*.

Invertebrata (Lat. *in*, not; *vertebra*, a joint of the backbone)—Animals which are destitute of a skull and spinal column.

Labium (Lat. lip)—The *lower* lip in the *Arthropoda*.

Labrum (Lat. lip)—The *upper* lip in the *Arthropoda*.

Labyrinthodon (Gr. *labyrinthos*, a maze; *odous*, tooth)—An extinct amphibian, so called from the structure of its teeth.

Lamellibranchiata (Lat. *lamella*, a plate; Gr. *branchia*, a gill)—A class of *Mollusca*, so called because their gills are composed of folds of membrane.

Larva (Lat. a mask)—The first stage of an insect after it comes from the egg. It is commonly called a caterpillar.

Lens (Lat. a bean)—A term applied to the hexagonal facets into which the eyes of insects are divided.

Lobes—Various *rounded* portions of animals are so called.

Lucernarida (Lat. *lucerna*, a lamp)—A group of *Hydrozoa*.

Lumbar (Lat. *lumbus*, the loin)—Belonging to the loins.

Madreporiform—Applied to the tubercle by which the ambulacral system of the sea-urchins communicates with the exterior. It is so called because it is perforated with small holes like madrepore coral.

Mammalia (Lat. *mamma*, the breast)—The class of *Vertebrata* which suckle their young.

Mandible (Lat. *mando*, I chew)—The lower jaw in the *Vertebrata*; in *Arthropoda*, the upper pair of jaws; the beak of the *Cephalopoda*.

Mantle—The outer covering of the *Mollusca*.

Manubrium (Lat. a handle)—The handle-shaped polypite which is suspended from the centre of the disc in the *Medusæ*.

GLOSSARY.

Marsupialia (Lat. *marsupium*, a pouch)—An order of *Mammalia* in which the immature young are protected in a pouch, placed beneath the abdomen of the female.

Masticatory (Lat. *mastico*, I chew)—A term applied to organs used in chewing food.

Maxillæ (Lat. jaws)—The lower pair, or pairs, of jaws in the *Arthropoda*.

Maxillipedes (Lat. *maxillæ*, jaws; *pes*, a foot)—Or "foot jaws," the modified limbs of the *Crustacea*, which are used as masticatory organs.

Medusæ—The sea-nettles or jelly-fishes, so called because their tentacles resemble the hair of the *Medusa*, or chief of the gorgons, which was said to consist of snakes.

Mesenteries (Gr. *mesos*, the middle; *enteron*, an intestine)—The vertical partitions which divide into chambers the intervening space between the alimentary tube and the body wall of a sea-anemone.

Metacarpus (Gr. *meta*, after; *carpus*, the wrist)—The bones which intervene between the wrist and the fingers in the higher vertebrates.

Metatarsus (Gr. *meta*; *tarsos*, the hollow part of the foot)—The bones which intervene between the tarsus or ankle bones and the phalanges of the toes.

Molars (Lat. *mola*, a mill)—The teeth which are not preceded by a milk set—the "grinders."

Mollusca (Lat. *mollis*, soft)—One of the sub-kingdoms, so called from the soft nature of their bodies. It includes the classes *Cephalopoda*, *Gasteropoda*, *Pteropoda*, and *Lamellibranchiata*.

Molluscoida (Lat. *mollis*, soft; Gr. *eidos*, form)—A sub-kingdom, including the *Brachiopoda*, *Ascidioida*, and *Polyzoa*.

Monad (Gr. *monas*, a unit)—An exceedingly minute infusorian.

Monodelphia (Gr. *monos*, single; *delphus*, womb)—A division of Mammalia, including all the higher members of the group.

Monotremata (Gr. *monos*, single; *trema*, an opening)—An order of *Mammalia*, in which the intestine and the ducts of the urinary and genital organs open into a common cloaca. It includes the duck mole and the porcupine ant-eater.

Multivalve (Lat. *multus*, many; *valvæ*, folding doors)—A term applied to certain gasteropod shells which are composed of several pieces.

Myriapoda (Gr. *murios*, ten thousand; *pous*, a foot)—A class of *Arthropoda*, distinguished by having numerous feet. It includes the centipedes and millipedes.

Nectocalyx (Gr. *necho*, I swim; *kalux*, a cup)—The swimming bell of a jelly-fish.

Nematoidea (Gr. *nema*, thread; *eidos*, form)—A group of *Scolecida*, including the "thread worms" and the "round worms."
Neural (Gr. *neuron*, a nerve)—Belonging to the nervous system.
Nucleated—Possessing a nucleus or central solid particle.
Nucleolus—A minute particle attached to the "nucleus," or supposed ovary of the *Infusoria*.

Occipital—Belonging to the *occiput*, or that part of the skull which forms the back portion of the head.
Ocelli (Lat. diminutive of *oculus*, eye)—The simple eyes of spiders, crustaceans, molluscs, and some insects.
Octopoda (Gr. *octo*, eight; *pous*, foot)—A group of cuttle-fishes, possessing eight arms.
Odontophore (Gr. *odous*, tooth; *phero*, I fear)—The lingual ribbon, or tooth-bearer, of the higher *Mollusca*.
Œsophagus (Gr. *oisos*, a reed; *phago*, I eat)—The tube leading from the mouth to the stomach.
Operculum (Lat. a lid)—The bony flap which covers the gills of fishes; the horny disc which closes the shells of the *Gasteropoda*.
Ornithodelphia (Gr. *ornis*, a bird; *delphus*, womb)—The lowest division of the *Mammalia*, including the *Monotremata*.
Oscula (Lat. diminutive of *os*, mouth)—The large or exhalant apertures of sponges.
Otoliths (Gr. *ous*, ear; *lithos*, stone)—Concretions found in the auditory sacs of fishes, *Crustacea*, and *Mollusca*.
Ovoviviparous (Lat. *ovum*, egg; *vivo*, I live; *pario*, I bring forth)—A term applied to those animals which retain the eggs within their bodies until they are hatched.

Pallium (Lat. a cloak)—The "mantle" of the Mollusca; "pallial line," the impression left in the shell by the muscular margin of the mantle.
Palpi (Lat. *palpo*, I touch)—Organs of touch connected with the appendages of the mouth in the *Arthropoda*.
Parietal (Lat. *paries*, a wall)—Belonging to the walls of the body.
Parieto-splanchnic (Lat. *paries*; Gr. *splanchna*, the internal organs)—A ganglion in the higher Mollusca, which sends nerve-fibres to the mantle, gills, and internal organs.
Pectoral (Lat. *pectus*, the chest)—Belonging to the chest.
Pedal (Lat. *pes*, foot)—Belonging to the foot, in *Mollusca*.
Pedicellariæ (Lat. *pedicellus*, a louse)—Curious appendages attached to the sea-urchins.
Peduncle (Lat. *pedunculus*, a stalk)—The muscular stalk by which the Brachiopoda are attached; the stem by which the barnacle connects itself with wood or other objects.

GLOSSARY.

Pelvis—The bony arch to which the hind limbs of the *Vertebrata* are attached.

Pericardium (Gr. *peri*, around; *kardia*, the heart)—The membrane that surrounds the heart.

Perigastric (Gr. *peri; gaster*, the stomach)—The space which surrounds the stomach and other internal organs.

Perivisceral (Gr. *peri;* Lat. *viscera*, the internal organs)—The space surrounding the viscera.

Phalanges (Gr. *phalanx*, a row)—The bones which compose the digits of the higher *Vertebrata*.

Pharynx—The upper part of the gullet.

Phragmacone (Gr. *phragma*, a partition; *konos*, a cone)—That part of the internal shell of a belemnite which is divided into chambers by partitions.

Pisces (Lat. *piscis*, a fish)—A class of *Vertebrata*.

Placenta (Lat. a cake)—An organ containing a network of blood-vessels, by which the young of the higher mammals receive nourishment from the mother before birth.

Placoid (Gr. *plax*, a plate; *eidos*, form)—Applied to the scales of sharks and rays, which are in the form of irregular bony plates, sometimes armed with spines.

Planarida (Gr. *plane*, wandering)—A group of *Turbellaria*.

Pleura (Gr. the side)—The membrane which surrounds the lungs of the higher *Vertebrata*.

Pneumatic (Gr. *pneuma*, air)—Containing air.

Polycystina (Gr. *polus*, many; *kustis*, a bladder)—A group of *Protozoa* with minute perforated siliceous shells.

Polygastrica (Gr. *polus; gaster*, stomach)—The name given by Ehrenberg to the *Infusoria*.

Polypary—The chitinous covering of the compound *Hydrozoa*.

Polype (Gr. *polus*, many; *pous*, foot)—A single individual in the *Actinozoa*.

Polypide—One of the zöoids in the *Polyzoa*.

Polypite—One of the zöoids in the *Hydrozoa*.

Polyzoa (Gr. *polus*, many; *zöon*, an animal)—A class of Molluscoida—called also *Bryozoa*.

Polyzoarium—The polypidum, or chitinous covering of the *Hydrozoa*.

Præmolars—The permanent molars which succeed the molars of the milk set of teeth; the bicuspid teeth.

Procœlus (Gr. *pro*, front; *koilos*, hollow)—Applied to vertebræ which are hollow in front.

Protoplasm (Gr. *protos*, first; *plasso*, I mould)—The primitive basis of all organic tissues, sometimes used as a synonym for *sarcode*.

Protozoa (Gr. *protos; zöon*, an animal)—The lowest of the sub-kingdoms.

Pseudo-hæmal (Gr. *pseudos*, false; *haima*, blood)—Applied to the circulatory system of the *Annelida*.

Pseudo-hearts—Contractile cavities found in the Brachiopoda, formerly believed to be hearts.

Pseudo-Navicellæ (Gr. *pseudos*; *Navicula*, a diatom)—The undeveloped form of the Gregarinida.

Pseudo-podia (Gr. *pseudos*; *pous*, foot)—Processes thrust out by the *Rhizopoda* from the surface of their bodies, which serve the purpose of limbs.

Pteropoda (Gr. *pteron*, wing; *pous*, foot)—A class of Mollusca which swim by means of wing-like processes attached to the head.

Pulmo-Gasteropoda—A group of Gasteropods which breathe by pulmonary sacs or lungs.

Pupa (Lat. a doll)—The stage of metamorphosis of an insect intermediate between the *larva* and *imago*. It is also called a chrysalis.

Radiata (Lat. *radius*, a ray)—One of the sub-kingdoms of Cuvier. This group is now broken up, its members being assigned to several sub-kingdoms (*Cœlenterata, Annuloida, Protozoa, &c.*)

Radiolaria (Lat. *radius*, a ray)—A class of *Protozoa*, including the *Polycystina*.

Radius—One of the two bones of the fore-arm of the higher vertebrates.

Ramus (Lat. a branch)—One of the halves of the lower jaw in the *Vertebrata*.

Reptilia (Lat. *repo*, I creep)—A class of *Vertebrata*, including crocodiles, lizards, serpents, and tortoises.

Rhizopoda (Gr. *rhiza*, root; *pous*, a foot)—A class of *Protozoa* which have the power of pushing out *pseudopodia*. These processes have the appearance of *roots*.

Rotifera (Lat. *rota*, wheel; *fero*, I carry)—A group of microscopic animals belonging to the class *Scolecida*. The mouth is surrounded by fringes of cilia, which, when in motion, resemble a pair of toothed wheels.

Rugosa (Lat. *rugosus*, wrinkled)—An extinct group of corals.

Sacrum—The vertebræ which articulate with the haunch bones to form the pelvis.

Sarcode (Gr. *sarx*, flesh; *eidos*, form)—The jelly-like substance of which the bodies of the *Protozoa* are formed. It is sometimes called *protoplasm*.

Scapula—The shoulder blade.

Sclerobasic (Gr. *scleros*, hard; *basis*, the foundation)—A term applied to the red coral, because it is secreted by the united *bases* of the polypes.

Sclerodermic (Gr. *scleros*, hard ; *derma*, skin)—Applied to the coral which is formed within the tissues of the reef-building polypes.

Scolecida (Gr. *skolex*, a worm)—A class of *Annuloida*, including the tape-worm and other internal parasites.

Septa (Lat. partitions)—Applied to the walls of the chambers in the nautilus, &c.

Sertularida (Lat. *sertum*, a wreath)—A group of *Hydrozoa*.

Sessile (Lat. *sedeo*, I sit)—Not supported upon stalks; the opposite of pedunculated.

Siliceous (Lat. *silex*, flint)—Composed of silex or flint.

Siphon (Gr. a tube)—One of the breathing tubes in the *Mollusca*.

Siphuncle (Lat. *siphunculus*, a little tube)—The tube which connects the chambers of the shells in some of the *Cephalopoda*.

Somatic (Gr. *soma*, the body)—Belonging to the body.

Somite (Gr. *soma*)—A segment of the body in the *Annulosa*.

Spermatozoa (Gr. *sperma*, seed; *zoon*, an animal)—Minute organisms found in the sperm cells of animals.

Spicula (Lat. *spiculum*, a point)—Applied to the needle-shaped, siliceous bodies found in sponges, &c.

Spinnerets—The teat-like organs situated at the extremity of the abdomen in spiders, &c., through which the threads that form the webs are drawn.

Spongida—A group of *Rhizopoda*, including the sponges.

Sternum—The breast-bone.

Stigmata (Gr. *stigma*, a mark)—The apertures by which the tracheæ of insects, &c., communicate with the atmosphere.

Swimmerets—Limbs attached to the abdomen in the *Crustacea* which are used in swimming.

Tæniada (Gr. *tainia*, a ribbon)—The group of *Scolecida* to which the tape-worms belong.

Tarso-metatarsus—A bone in the leg of a bird formed by the amalgamation of the lower portion of the tarsus with the metatarsus.

Tarsus—The small bones which form the ankle.

Telson (Gr. the end)—The somite at the extremity of the abdomen in the *Crustacea*.

Test (Lat. *testa*, a shell)—Applied to the calcareous covering of the sea-urchins, &c.

Tetrabranchiata (Gr. *tetras*, four ; *branchiæ*, gills)—An order of *Cephalopoda* to which the pearly nautilus belongs. They possess four gills.

Thalassa-collida (Gr. *thalassa*, the sea; *kolla*, glue)—A group of *Protozoa*.

Theca (Gr. *theke*)—A sheath or case.

Thorax (Gr. a breastplate)—The chest.

Tibia (Lat. a flute)—One of the two bones of the leg.

Trachea (Gr. *tracheia*, the windpipe)—The tube which connects the lungs with the mouth in the air-breathing *Vertebrata*.

Tracheæ—The air-tubes which ramify the bodies of insects, &c.

Trematoda (Gr. *trema*, a pore)—A group of *Scolecida* to which the "liver-fluke" of the sheep belongs.

Trilobites (Gr. *treis*, three; *lobos*, a lobe)—A group of extinct *Crustacea* found abundantly in the Silurian rocks. They are so called because their bodies were composed of three lobes.

Tunicata (Lat. *tunica*, a cloak)—A class of *Molluscoida* which are covered with a leathery case. It is also called *Ascidioida*.

Turbellaria (Lat. *turbo*, I disturb)—A group of free-swimming *Scolecida*. They are so called on account of the currents produced by the cilia with which their bodies are covered.

Ulna (Gr. *olene*, the elbow)—One of the bones of the fore-arm.

Umbo (Lat. a boss)—The beak of a bivalve shell.

Univalve (Lat. *unus*, one; *valvæ*, folding doors)—A shell composed of a single piece. Applied to the shells of the *Cephalopoda*, and most of the *Gasteropoda*.

Vascular (Lat. *vas*, a vessel)—Belonging to the circulatory system.

Velum (Lat. a veil)—The membrane which partly closes the disc in the "naked-eyed" *medusæ*.

Ventral (Lat. *venter*, the stomach)—Belonging to the lower surface of the body.

Ventricle (Lat. *venter*)—One of the cavities of the heart which receives blood from the auricle, and drives it either to the breathing organs or through the system.

Vertebra (Lat. *verto*, I turn)—One of the joints of the back-bone.

Vertebrata—The highest sub-kingdom, characterized by the possession of a back-bone.

Vesicle (Lat. *vesica*, a bladder)—A little sac or bladder.

Viviparous (Lat. *vivo*, I live; *pario*, I bring forth)—Producing the young alive.

Zöoid (Gr. *zöon*, animal; *eidos*, form)—Applied to the individuals which make up a compound organism.

Zöophyte (Gr. *zöon*; *phuton*, a plant)—A term sometimes applied to animals which resemble plants, such as the corals, sea-anemones, sponges, &c.

www.ingramcontent.com/pod-product-compliance
Lightning Source LLC
Chambersburg PA
CBHW022115160426
43197CB00009B/1031